Electricity for Refrigeration, Heating, and Air Conditioning

Electricity for Refrigeration, Heating, and Air Conditioning Fourth Edition

Russell E. Smith
Athens Area Technical School

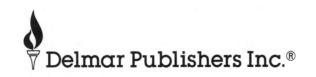

Delmar Publishers Inc.®

NOTICE TO THE READER

Delmar Staff
Executive Editor: Michael McDermott
Project Editor: Mary P. Robinson
Production Coordinator: Wendy A. Troeger
Design Coordinator: Karen Kunz Kemp

Cover Photo Credits:
Compressor photo courtesy of Copeland Corporation
Meter photo courtesy of John Fluke Mfg. Co., Inc.
Thermostat photo: Mike Gallitelli, Photographer

For information, address Delmar Publishers Inc.
3 Columbia Circle, PO Box 15015
Albany, New York 12212−5015

COPYRIGHT © 1992
BY DELMAR PUBLISHERS INC.

Printed in the United States of America
Published simultaneously in Canada
by Nelson Canada,
a division of The Thomson Corporation

10 9 8 7 6 5 4

Library of Congress Cataloging-in-Publication Data

Smith, Russell E., 1944-
   Electricity for refrigeration, heating, and air conditioning/
Russell E. Smith.—4th ed.
      p. cm.
   Includes index.
   ISBN 0-8273-4858-4 (text).—ISBN 0-8273-4859-2 (IG)
   1. Electric engineering.   2. Heating.   3. Air conditioning.
   4. Refrigeration and refrigerating machinery.   I. Title.
TK153.S57 1992   91-45579
621.3'024697—dc20   CIP

Table of Contents

Preface

Electricity for Refrigeration, Heating, and Air Conditioning initially was written because there was no text that adequately covered the electrical principles and practices required of an installation or service technician in the refrigeration, heating, and air conditioning industry. This text is written with a blend of theory and practice suitable for the vocational-technical student or the industry practitioner who wishes to upgrade his or her background.

The purpose of this text is to assemble concepts and procedures that will enable the reader to work successfully in the industry. Electrical principles, components, meters, schematic diagrams, and systems are explained and applied to small and large installations and equipment. Installation procedures are presented to enable personnel to determine the correct type and size of the conductor and to make the necessary connections in both electrical panels and equipment. Troubleshooting and servicing are presented using a practical approach in hopes of ensuring immediate productivity. Step-by-step procedures for reading schematic diagrams are intended to make reading the diagrams easier. This text does not replace manufacturers' service manuals but complements them.

In this fourth edition additional information has been provided in response to reviews and recommendations of instructors who teach the subject matter. Among the changes include rearrangement of the more difficult diagrams in later chapters; updated diagrams; reordering of some material to a more practical sequence; and dividing chapters for simplicity.

Electrical components used in the industry are constantly changing and improvements are being made. This edition attempts to cover the new electrical components that have made their way into the field since the previous edition. In addition, no new edition would be complete without additional troubleshooting information.

ACKNOWLEDGMENTS

Special thanks are to be given to the reviewers of this edition: Ronald Coffee, Mid America AJTS, Wayne, OK; Lloyd Cossey, Ozarka Vo Tech, Melbourne, AK; Rosser S. Farley III, Dundalk Community College, Baltimore, MD; Arthur Gibson, Erwen Vo-Tech Center, Tampa, FL; James Hamilton, College of the Desert, Palm Desert, CA; Greg Jourdan, Wenatchee Valley College, Wenatchee, WA; and Henry Michalski, College of DuPage, Glen Ellyn, IL.

The author would like to express sincere appreciation to the following companies, which generously supplied illustrations, technical data, or other resources and assistance during the preparation of this book.

Acme Electric Corporation
Airserco Manufacturing Company
Airtemp Corporation, Subsidiary of
Fedders Corporation
Allen-Bradley Company
Amana Refrigeration, Inc.
Amprobe Instrument
A.W. Sperry Instrument Co.
Barber Coleman Company
Carrier Corporation, Subsidiary of
United Technologies Corporation
The Chase Shawmut Company
Continental Wire & Cable Division of
the Anaconda Company
Copeland Corporation
Crouse-Hinos Company, Distribution
Equipment Division
Delco Products Division, General Motors
Corporation
Eagle Electric Manufacturing Company,
Inc.
Emerson Electric Company, Emerson
Motor Division
Essex Group, Controls Division
Fedders Corporation
Furnas Electric Company
General Electric Company, Appliance
Control Products Department
General Electric Company, Central Air
Conditioning Department
General Electric Company, Circuit
Protective Devices Products
Department
Gould, Inc., Electric Motor Division
Gould I-T-E Electrical Products
GTE Sylvania, Inc.
Heinemann Electric Company
Honeywell, Inc.
Imperial Clevite Inc., Fluid Components
Division

ITT General Controls
Jackes-Evans Manufacturing Company
John Fluke Manufacturing Co., Inc.
John L. Underwood Company, Inc.
Johnson Controls, Inc.
Knopp Inc.
Lennox Industries, Inc.
Motors & Armatures, Inc.
National Fire Protection Association
Paragon Electric Company, Inc., Subsidi-
ary of AMF Inc.
Penn Division, Johnson Controls, Inc.
Potter & Brumfield Division, AMF Inc.
Ranco Controls Division
Reliance Electric Company
Rheem Air Conditioning Division
Robbins & Myers, Electric Motor
Division
Robertshaw Controls Company
Robinair Manufacturing
Sealed Units Parts Company, Inc.
Simpson Electric Company, Elgin, Ill.
The Singer Company, Climate Control
Division
Sprague Electric Company
Square D Company
Tappan Air Conditioning Division
Tecumseh Products Company
Texas Instruments, Incorporated
TIF Instruments, Inc.
The Trane Company
Triplett Corporation
Universal Electric Company
Watsco, Inc.
Westinghouse Electric Corporation, Cen-
tral Residential Air Conditioning
Division
White-Rodgers Division, Emerson Elec-
tric Company
York Heating and Air Conditioning

Once again, thanks to my wife for her patience and help.

CHAPTER 1

Basic Electricity

INTRODUCTION

It would be extremely difficult for anyone to be unaware of the importance of electricity to modern society. In any direction a person looks, electricity can be seen at work, supplying the needs and luxuries of modern life. Electricity allows people to use light bulbs for light, motors to power rotating machinery, and electric heaters for heat. Electricity is one of the most important technological advances in the history of the world.

Without electricity the heating, cooling, and refrigeration industry would never have advanced beyond its infancy. The industry relies almost totally on electricity for the operation of any heating, cooling, or refrigeration system.

Most control systems used in the heating, cooling, and refrigeration industry use electrical energy to operate the control system, maintaining the desired temperature. Electrical components in systems that require rotation, such as compressors and fan motors, use electric motors to accomplish this rotation. Many other devices such as electric heaters, solenoid valves, and signal lights that are incorporated in equipment also require electrical energy for operation. The use of electricity can be seen in all aspects of the industry.

Along with all the electric devices used in systems today come problems that are, in most cases, electrical and that must be corrected by field service personnel. Thus, it is essential for all industry personnel to understand the basic principles of electricity so that they are able to perform their jobs in the industry.

We begin our study of electricity with a discussion of atomic structure.

1.1 ATOMIC THEORY

Matter is the substance of which a physical object is composed, whether it be a piece of iron, wood, or cloth, or whether it is a gas, liquid, or solid.

Matter is composed of fundamental substances called **elements**. There are 105 elements that have been found in the universe. Elements, in turn, are composed of atoms. An **atom** is the smallest particle of an element that can exist alone or in combination. All matter is made up of atoms or a combination of atoms, and all atoms are electrical in structure.

Suppose a piece of chalk is broken in half and one piece discarded. Then the remaining piece is broken in half and one piece discarded. If this procedure is continued, eventually the piece of chalk will be broken into such a small piece that by breaking it once more there will no longer be a piece of chalk but only a molecule of chalk. A **molecule** is the smallest particle of a substance that has the properties of that substance. If a molecule of chalk is broken down into smaller segments, only individual atoms will exist, and they will no longer have the properties of chalk. The atom is the basic building block of all matter. The atom is the smallest particle that can combine with other atoms to form molecules.

Although the atom is a very small particle, it is also composed of several parts. The central part is called the **nucleus**. Other parts, called **electrons**, orbit around the nucleus. Each electron is a relatively small, negatively charged particle. The electrons orbit the nucleus in much the same way that the planets orbit the sun.

The nucleus, the center section of an atom, is composed of protons and neutrons. The **proton** is a heavy, positively charged particle. The proton has an electric charge that is opposite but equal to that of the electron. All atoms contain a like number of protons and electrons. The **neutron** is a neutral particle, which means that it is neither positively nor negatively charged. The neutrons tend to hold the protons together in the nucleus.

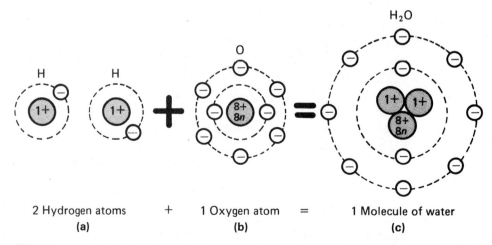

2 Hydrogen atoms   +   1 Oxygen atom   =   1 Molecule of water
(a)   (b)   (c)

FIGURE 1.1 Atomic structure of a water molecule (one atom of oxygen and two atoms of hydrogen)

The simplest atom that exists is the hydrogen atom, which consists of one proton that is orbited by one electron, as shown in Figure 1.1(a). All atoms are not as simple as the hydrogen atom. Other atoms have more particles. The difference in each different atom is the number of electrons, neutrons, and protons that the atom contains. The hydrogen atom has one proton and one electron. The oxygen atom has 8 protons, 8 neutrons, and 8 electrons, as shown in Figure 1.1(b). The silver atom contains 47 protons, 61 neutrons, and 47 electrons. The more particles an atom has, the heavier the atom is. Since there are 105 elements, but millions of different types of substances, there must be some way of combining atoms and elements to form these substances.

When elements (and atoms) are combined, they form a chemical union that results in a new substance, called a **compound**. For example, when two hydrogen atoms combine with one oxygen atom, the compound water is formed. The atomic structure of one molecule of water is shown in Figure 1.1(c).

The chemical symbol for a compound denotes the atoms that make up that compound. Refrigerant 12 is a substance commonly used in refrigeration systems. A refrigerant is a fluid that absorbs heat inside the conditioned area and releases heat outside the conditioned area. The chemical symbol for one molecule of refrigerant 12 is CCl_2F_2. One molecule of the refrigerant contains one atom of carbon, two atoms of fluorine, and two atoms of chlorine. The chemical name for refrigerant 12 is dichlorodifluoromethane. All materials can be identified according to their chemical makeup, that is, the atoms that form their molecules.

1.2 POSITIVE AND NEGATIVE CHARGES

An atom usually has an equal number of protons and electrons. When this condition exists, the atom is electrically neutral because the positively charged protons exactly balance with the negatively charged electrons. However, under certain conditions an atom can become unbalanced by losing or gaining an electron. When an atom loses or gains an electron, it is no longer neutral. It is either negatively or positively charged, depending on whether the electron is gained or lost. Thus, in an atom a charge exists when the number of protons and electrons is not equal.

Under certain conditions some atoms can lose a few electrons for short periods. Electrons that are in the outer orbits of some materials, especially metals, can be easily knocked out of their orbits. Such electrons are referred to as **free electrons**, and materials with free electrons are called **conductors**. When electrons are removed from the atom, the atom becomes positively charged because the negatively charged electrons have been removed, creating an unbalanced condition in the atom.

An atom can just as easily acquire additional electrons. When this occurs, the atom becomes negatively charged.

Charges are thus created when there is an excess of electrons or protons in an atom. When one atom is charged and there is an unlike charge in another atom, electrons can flow between the two. This electron flow is called **electricity**.

An atom that has lost or gained an electron is considered unstable. A surplus of electrons in an atom creates a negative charge. A shortage of electrons creates a positive charge. Electric charges react to each other in different ways. Two negatively charged particles repel each other. Positively charged particles also repel each other. Two opposite charges attract each other. The **law of electric charges** states this: Like charges repel and unlike charges attract. Figure 1.2 shows an illustration of the law of electric charges.

All atoms tend to remain neutral because the outer orbits of electrons repel other electrons. However, many materials can be made to acquire a positive or negative charge by some mechanical means, such as friction. The familiar crackling when a hard rubber comb is run through hair on a dry winter day is an example of an electric charge generated by friction.

1.3 FLOW OF ELECTRONS

The flow of electrons can be accomplished by several different means: friction, which produces static electricity; chemical, which produces electricity in a battery; and magnetic (induction), which produces electricity in a generator. Other methods are also used, but the three mentioned here are the most common.

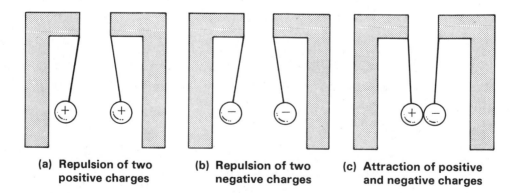

(a) Repulsion of two positive charges

(b) Repulsion of two negative charges

(c) Attraction of positive and negative charges

FIGURE 1.2 Like charges repel and unlike charges attract each other.

Static Electricity

The oldest method of moving electrons is by **static electricity.** Static electricity produces a flow of electrons by permanently displacing an electron from an atom. The main characteristic of static electricity is that a prolonged or steady flow of current is not possible. As soon as the charges between the two substances are equalized (balanced), electron flow stops.

Friction is usually the cause of static electricity. Sliding on a plastic seat cover in cold weather and rubbing silk cloth on a glass rod are two examples of static electricity produced by friction. Static electricity, no matter what the cause, is merely the permanent displacement or transfer of electrons. To obtain useful work from electricity, a constant and steady flow of electrons must be produced.

Electricity Through Chemical Means

Electricity can also be produced by the movement of electrons due to chemical means. A battery produces an electron flow by a chemical reaction that causes a transfer of electrons between two **electrodes**. One electrode collects electrons and one gives away electrons. The dry cell battery uses two electrodes made of two dissimilar metals inserted in a pastelike **electrolyte**. Electricity is produced when a chemical reaction occurs in the electrolyte between the electrodes, causing an electron flow. The construction of a dry cell battery is shown in Figure 1.3.

The container of a dry cell battery, which is made of zinc, is the negative electrode (gives away electrons). The carbon rod in the center of the dry cell is the positive electrode (collects electrons). The space between the electrodes is filled with an electrolyte, usually manganese dioxide paste. The acid paste causes a chemical reaction between the carbon electrode and the zinc case. This reaction displaces the electrons, causing an electrode flow. The top of the dry cell is sealed to prevent the electrolyte from drying and to allow the cell to be used in any position. The dry cell battery will eventually lose all its power because energy is being used and not being replaced.

The storage battery is different from a dry cell battery because it can be recharged. Thus, it lasts somewhat longer than a dry cell battery. But it, too, will eventually lose all its energy.

The storage battery consists of a liquid electrolyte and negative and positive electrodes. The electrolyte is diluted sulfuric acid. The positive electrode is coated with lead dioxide and the negative electrode is sponge lead. The chemical reaction between the two electrodes and the electrolyte displaces electrons and creates voltage between the plates. The storage battery is

FIGURE 1.3 Construction of a dry cell battery

FIGURE 1.4 Common storage battery used in automobile electric systems

recharged by reversing the current flow into the battery. The storage battery shown in Figure 1.4 is commonly used in automobile electric systems.

Electricity Through Magnetism

The magnetic or induction method of producing electron flow uses a conductor to cut through a magnetic field, which causes a displacement of electrons. The alternator, generator, and transformer are the best examples of the magnetic method. The magnetic method is used to supply electricity to consumers.

The flow of electrons in a circuit produces magnetism, which is used to cause movement, or thermal energy, which in turn is used to cause heat. A magnetic field is created around a conductor—an apparatus for electrons to flow through—when there is a flow of electrons in the conductor. The flow of electrons through a conductor with a resistance will cause heat, such as in an electric heater.

The heating, cooling, and refrigeration industry uses magnetism to close relays and valves and to operate motors by using coils of wire to increase the strength of the magnetic field.

1.4 CONDUCTORS AND INSULATORS

The structure of an atom of an element is what makes it different from the atom of another element. The number of protons, neutrons, and electrons and the arrangement of the electrons in their orbits vary from element to element. In some elements, the outer electrons rotating around the nucleus are easily removed from their orbits. As stated earlier, elements that have atoms with this characteristic are called **conductors**. A conductor can transmit electricity or electrons.

Most metals are conductors, but all metals do not conduct electricity equally well. The best conductors are silver, copper, and aluminum. The high cost of silver prevents it from being used widely. Its use is largely limited to contacts in certain electrical switching devices such as contactors and relays. Copper, almost as good a conductor as silver, is usually used because it is less expensive.

Materials that do not easily give up or take on electrons are called **insulators**. An insulator retards the flow of electrons. Glass, rubber, and asbestos are examples of insulators. How well an insulator prevents electron flow depends on the strength of that flow. If the flow of electrons is strong enough, the insulator will break down, causing electrons to flow through it.

There is no perfect insulator. All insulators will break down under certain conditions if the electron flow is high enough. Increasing the thickness of the insulation helps overcome this problem.

Conductors and insulators are important parts of electric circuits and electric systems. They are widely used in all electric components in the industry.

1.5 ELECTRIC POTENTIAL

In a water system, water can flow as long as pressure is applied to one end of a pipe and the other end of the pipe is open. The greater the pressure in a water system, the greater the quantity of water that will flow. Similarly, in an electrical system electrons will flow as long as electric pressure is applied to the system. **Voltage**, **potential difference**, and **electromotive force** are all terms used to describe electric pressure.

Recall that the law of electric charges states that unlike charges attract. Consequently there is a pull, or *force*, of attraction between two dissimilarly charged objects. We call this pull of attraction a **field of force**.

Another way of looking at this is to picture excess electrons (the negative charge) as straining to reach the point where there are not enough electrons (the positive charge). If the two charges are connected by a conductor, the excess electrons will flow to the point where there are not enough electrons. But if the two charges are separated by an insulator, which prevents the flow of electrons, the excess electrons cannot move. Hence, an excess of electrons will pile up at one end of the insulator, with a corresponding lack, or deficiency, of electrons at the other end.

As long as the electrons cannot flow, the field of force between the two dissimilarly charged ends of the insulator increases. The resulting strain between the two ends is called the **electric pressure**. This pressure can become quite great. After a certain limit is reached, the insulator can no longer hold back the excess electrons, as discussed in the previous section. Hence, the electrons will rush across the insulator to the other end.

Electric pressure that causes electrons to flow is called **voltage**. Voltage is the difference in electric potential (or electric charge) between two points. The **volt** (V) is the amount of pressure required to force one **ampere** (A, the unit of measurement for current flow) through a resistance of one **ohm** (Ω, the unit of measurement for resistance; Ω is the Greek letter omega). In the industry, voltage is almost always measured in the range of the common volt. In other areas, the volt may be measured on a smaller scale of millivolt (mV), or one-thousandth of a volt. For larger measurements of the volt, the kilovolt (kV), equal to 1000 volts, is used.

$$1 \text{ millivolt} = 0.001 \text{ volt}$$
$$1 \text{ kilovolt} = 1000 \text{ volts}$$

To maintain electric pressure we must have some means to move electrons in the same manner that water pressure moves water. In an electric circuit, this can be maintained by a battery, as shown in Figure 1.5, or by a generator or alternator, as shown in Figure 1.6. The battery forces electrons to flow to the positive electrode and causes electric pressure. A generator causes electric pressure by transferring electrons from one place to another.

FIGURE 1.5 A dry cell battery supplying electric potential (voltage) to an electric circuit

FIGURE 1.6 A generator supplying electric potential (voltage) to an electric circuit

Electromotive force can be produced in several ways. The easiest method to understand is the simple dry cell battery discussed in Section 1.3. The most popular method of producing an electromotive force is by using an alternating current generator. The alternating current generator is supplied with power from another source. Then a wire loop is rotated through the magnetic field created by the voltage being applied, and an electromotive force is produced through the wire loop. We will discuss these ideas in more detail in succeeding sections.

1.6 CURRENT FLOW

Electrons flowing in an electric circuit are called **current**. Current flow can be obtained in an electric circuit by a bolt of lightning, by static electricity, or by electron flow from a generator. Figure 1.7 shows an electric system with electric pressure; the quantity of electrons flowing is also given.

There are two types of electric current: **direct current** and **alternating current**. Direct current flows in one direction. It is the type of current produced by dry cell batteries. Direct current is rarely used in the industry as a main power source but is used in some modern control circuits.

Alternating current flows back and forth. It is the type of current supplied to most homes by electric utility companies. It is the most commonly used source of electric potential in the heating, cooling, and refrigeration industry. Alternating current will be discussed in more detail in Chapter 5.

The current in an electric circuit is measured in **amperes (A)**. An ampere is the amount of current required to flow through a resistance of one ohm with a pressure of one volt. An ampere is measured with an **ammeter**. In the industry, the ampere is used almost exclusively. If a smaller unit of ampere measurement is required, the milliampere (mA), which is one-thousandth of an ampere, can be used. For larger measurements of amperes, the kiloampere (kA) can be used. One kiloampere equals 1000 amperes.

FIGURE 1.7 An electric system; electric potential forces electrons through a wire (a conductor).

1 milliampere = 0.001 ampere
1 kiloampere = 1000 amperes

The current that an electric device consumes can be used as a guide to the correct operation of the equipment by installation and service personnel. The electric motor is the largest current-consuming device in most heating, cooling, and refrigeration systems. The larger the electric device (load), the larger the current flow. Any electric device that uses electricity requires a certain current when operating properly.

1.7 RESISTANCE

Resistance is opposition to the flow of electrons in an electric circuit and is measured in ohms. Figure 1.8 shows two electric systems with different resistances. One ohm is the amount of resistance that will allow one ampere to flow with a pressure of one volt. The industry uses simple ohms for resistance in most cases because the scale is broad enough for most applications. In some special cases, the microhm (μΩ), which is one-millionth of an ohm, is used for extremely small resistance readings. Larger resistance readings are read in megohms (MΩ); one megohm is equal to one million ohms.

1 microhm = 0.000001 ohm
1 megohm = 1,000,000 ohms

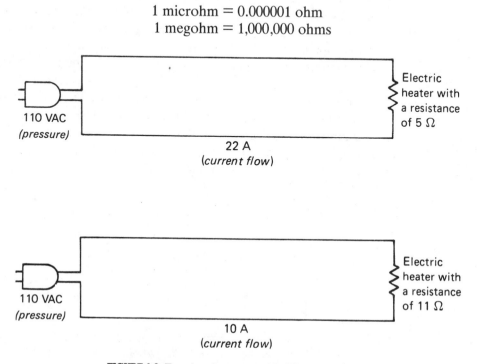

FIGURE 1.8 Two electric systems with different resistances

All electric devices will have a certain resistance. That resistance depends on the size and purpose of the device. As service personnel, you will have to become familiar with this value in components. If the resistance deviates far from the specified value or the estimated value, the device can be considered faulty.

1.8 ELECTRIC POWER AND ENERGY

When electrons move from the negative to the positive end of a conductor, work is done. **Electric power** is the rate at which the electrons do work. That is, electric power is the rate at which electricity is being used. The power of an electric circuit is measured in **watts (W)**. A watt of electricity is one ampere flowing with a pressure of one volt. The electric power of a circuit is the voltage times the amperage.

electric power = voltage × amperage

In an alternating current circuit, the voltage and current are not in phase. To obtain the correct power consumed by the circuit, the product of the voltage times the amperage must also be multiplied by a **power factor**. The power factor is the actual power divided by the calculated power and is expressed as a percentage. In a direct current circuit, the product of the voltage times the amperage gives the power of the circuit; the power factor is not needed.

The industry uses the units of watts for devices that consume a small amount of power. Examples of such devices are small electric motors and small resistance heaters. Other units used are the horsepower (hp) and the British thermal unit (Btu). One horsepower is equal to 746 watts. One watt is equal to 3.41 Btu per hour.

1 horsepower = 746 watts
1 watt = 3.41 Btu/hour

These conversion figures are often used in the industry to calculate the Btu rating of an electric heater if the watt rating is known. The horsepower conversion is used to calculate the horsepower of a motor only if the watts are known.

The rate at which electric power is being used at a specific time is called **electric energy**. Electric energy is measured in watthours (Wh). For example, a one-horsepower motor uses 746 watts in one hour. It therefore uses 746 watthours.

The wattage rating or consumption of any electric device only denotes the amount of power the device is using. However, time must be considered when calculating electric energy, that is, the power being consumed over a definite period. Watthours give the number of watts used for a specific period of time.

The units of kilowatts (kW) are usually used to determine the amount of electricity consumed. Thus, an electric utility calculates the power bill of its customers using **kilowatthours** (kWh) because the watthour readings would be extremely large. The kilowatthour reading is relatively small. One thousand watts used for one hour equals one kilowatthour. All electric meters used to measure the consumption of electricity record consumption in kilowatthours.

Heating and air-conditioning personnel are often required to make a calculation for the output of an electric heater in Btu's rather than in watts. The industry rates electric heating equipment in watts or kilowatts, which does not give consumers figures that they can understand. Most consumers are familiar with the term Btu and know basically what it means in terms of heat output. Therefore, the consumer often will request the Btu output rather than the wattage of a heating appliance. The Btu output can be easily calculated by multiplying the number of watts by the conversion factor of 3.41 Btu/h. Figure 1.9 shows an electric data plate for an electric heater. The wattage on the data plate is 4800 watts. Therefore the Btu output is

$$4800 \times 3.41 \text{ Btu/h} = 16,368 \text{ Btu/h/stage}$$

Another term that is used because of the high cost of energy is the **seasonal energy efficiency rating** (SEER) of an air conditioning unit or heat pump,

FIGURE 1.9 Data plate of an electric heater

measured in Btu's per watt. The SEER is the Btu output of the equipment divided by the power input with a seasonal adjustment. For example, if an air-conditioning unit has a SEER of 10, it will produce 10 Btu of cooling per watt of power consumed by the equipment. All air-conditioning manufacturers use SEER ratings for their equipment. The standards for SEER are set by the Air Conditioning and Refrigeration Institute.

1.9 OHM'S LAW

The relationship among the current, electromotive force, and resistance in an electric circuit is known as Ohm's law. In the nineteenth century, George Ohm developed the mathematical comparisons of the major factors in an electric circuit. Stated in simple terms, Ohm's law says that the greater the voltage, the greater the current, and the greater the resistance, the lesser the current. Ohm's law is represented mathematically as "current is equal to the electromotive force divided by the resistance." The following equation expresses Ohm's law:

$$I = \frac{E}{R}$$

In the equation, I represents the current in amperes, E represents the electromotive force in volts, and R represents the resistance in ohms. Ohm's law can also be expressed by the following two formulas:

$$E = IR \qquad R = \frac{E}{I}$$

In any of the three formulas, when two elements of an electric circuit are known, the unknown factor can be calculated. Ohm's law does not apply to alternating current circuits because coils of wire produce different effects on alternating current. Alternating current will be discussed in a later chapter. However, the general ideas of Ohm's law apply to alternating current circuits.

The following examples show the relationships of the voltage, current, and resistance in an electric circuit.

Example 1   What is the current in the circuit shown in Figure 1.10?

Step 1: $I = \frac{E}{R}$

Step 2: $I = \frac{120}{10}$

Step 3: $I = 12$ A

FIGURE 1.10 Simple electric circuit for Example 1

Example 2   What is the resistance of a 100-watt light bulb if the voltage is 120 volts and the current is .83 amperes?

Step 1: $R = \dfrac{E}{I}$

Step 2: $R = \dfrac{120}{.83}$

Step 3: $R = 145 \ \Omega$

Example 3   What is the voltage supplied to the circuit in Figure 1.11?

Step 1: $E = IR$

Step 2: $E = 5 \times 48$

Step 3: $E = 240 \text{ V}$

FIGURE 1.11 Simple electric circuit for Example 3

Ohm's law allows the calculation of the missing factor if the other two factors are known or can be measured. Figure 1.12 shows a simple method for remembering Ohm's law. If one of the factors in the circle is covered, the letters remaining in the circle give the correct formula for calculating the covered factor.

1.10 CALCULATING ELECTRIC POWER

Electric power can be calculated by using the formula $P = IE$. Two other formulas can be used to calculate the electric power of an electric circuit by

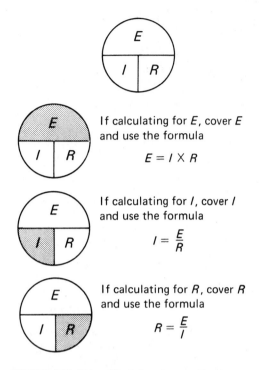

If calculating for E, cover E and use the formula

$$E = I \times R$$

If calculating for I, cover I and use the formula

$$I = \frac{E}{R}$$

If calculating for R, cover R and use the formula

$$R = \frac{E}{I}$$

FIGURE 1.12 Using Ohm's law (not applicable on AC inductive circuits)

substituting in the following equations:

$$P = \frac{E^2}{R} \quad P = I^2 R$$

The letter designations in these formulas are the same as in Ohm's law, with P representing power in watts.

The following three examples show the electric power calculations of three electric circuits.

Example 4   What is the power consumption of an electric circuit using 15 amperes and 120 volts?

Step 1: $P = IE$

Step 2: $P = 15 \times 120$

Step 3: $P = 1800$ W

Example 5   What is the current of an electric heater rated at 5000 watts on 230 volts?

Step 1: $I = \dfrac{P}{E}$

Step 2: $I = \dfrac{5000}{230}$

Step 3: $I = 21.7$ A

Example 6   What is the power of an electric circuit with 5 amperes current and 10 ohms resistance?

Step 1: $P = I^2 R$

Step 2: $P = 5^2 \times 10$

Step 3: $P = 25 \times 10$

Step 4: $P = 250$ W

SUMMARY

Everything—solids, liquids, and gases—is composed of matter. Matter can be broken down into molecules (the smallest particles of physical objects) and atoms (the smallest particles of an element that can exist alone or in combination). An atom is composed of a nucleus (the central part) and electrons (negatively charged) that orbit around the nucleus, much like the planets orbit the sun. The nucleus is composed of protons (positively charged) and neutrons (no charge). The number of protons is usually equal to the number of electrons, making the atom electrically neutral. When an atom loses electrons, it becomes positively charged. When it gains electrons, it becomes negatively charged. The law of charges states that like charges repel and unlike charges attract. Materials can be made to acquire positive or negative charges.

Electrons can be made to flow by use of friction, chemicals, and magnetism. A conductor is a material capable of transmitting electrons or electricity. Most metals are conductors. An insulator is a material that resists or prevents electron flow.

There are four important factors in any electric circuit: electromotive force, current, resistance, and power. The electromotive force of an electric circuit is the actual pressure in the circuit, much like water pressure in a water system. The electromotive force in an electric circuit is measured in volts. The voltage (pressure) must be sufficient to overcome the resistance of the circuit. Alternating current is used almost exclusively in the industry to supply electric power to equipment.

The number of electrons flowing in an electric circuit is called the current flow. The current flow of an electric circuit is measured in amperes. An ampere is the amount of current that will flow through a resistance of one ohm with a pressure of one volt.

The resistance of an electric circuit is measured in ohms. All electric loads have some resistance. Electric power is the rate at which electric energy is being used in an electric circuit. Electric power is measured in watts and kilowatts. The electric utilities use kilowatthours in most cases to charge their customers for the electric energy they have consumed. The kilowatthour is a measure of electric energy and takes into consideration the amount of time and the power consumption. One thousand watts used for a period of one hour equals one kilowatthour. Voltage, amperage, resistance, and wattage often use the prefixes kilo- or milli- to represent larger or smaller quantities of these factors and to avoid the use of extremely large or small numbers.

Ohm's law gives the relationship among the current, electromotive force, and resistance in an electric circuit. Ohm's law states the relationship mathematically. When any two factors in an electric circuit are known or can be measured, the formulas for Ohm's law can be used to find the third factor. Electric power can be calculated by using the formula $P = IE$.

QUESTIONS

1. What is an atom? What are the parts in an atom?

2. What is static electricity?

3. Name three ways electricity can be produced.

4. What parts do protons and electrons play in the production of electricity?

5. The simplest atom that exists is ____.

6. What are the three most important factors in an electric circuit?

7. What is electromotive force, and how is it measured?

8. What is current, and how is it measured?

9. What is resistance, and how is it measured?

10. What is electric power, and how is it measured?

11. Where do electrons exist in an atom, and what is their charge?

12. True or false. All atoms tend to lose electrons.

13. State the law of charges.

14. What is a proton? Where does it normally exist in an atom, and what is its charge?

15. Describe briefly the way a dry cell battery operates.

16. What is a conductor?

17. What is an insulator?

18. Why do metals make the best conductors?

19. In a battery the transfer of electrons occurs between two _____.

20. How do electric utilities charge consumers for electricity?

21. What is the source of electric pressure in an electric circuit?

22. What is the meaning of SEER when used in conjunction with an air conditioning unit?

23. State Ohm's law.

24. Why does Ohm's law not apply to alternating current?

25. What is the ampere draw of a 5000-watt electric heater used on 110 volts?

26. What is the resistance of the heating element of an electric iron if the ampere draw is 8 amperes when 115 volts are applied?

27. What is the voltage applied to a small electric heater if the heater is pulling 12 amperes and has a resistance of 10 ohms?

28. What is the Btu/h output of an electric heater rated at 15 kW?

29. What is the kilowatt output of an electric heater that has an ampere draw of 50 A and a voltage source of (a) 208 V? (b) 230 V?

2

Electric Circuits

INTRODUCTION

The electrical circuitry in a modern heating, cooling, and refrigeration system is important to the personnel who install or work on the electrical systems. Any electrical system used in the industry is composed of various types of circuits. Each type is designed to do a specific task within the system. We will look at some commonly used types of circuits in this chapter.

The two most important kinds of circuits are parallel circuits and series circuits. A **parallel circuit** is an electric circuit that has more than one path through which electricity may flow. A parallel circuit is designed to supply more than one load in the system.

A **series circuit** is an electric circuit that has only one path through which electricity may flow. It is usually used for devices that are connected in the circuit for safety or control.

A series-parallel circuit is an electrical circuit that has a combination of series and parallel circuits. Most electrical systems in equipment or control systems are made up of combination circuits. Electrical loads that require electrical power for operation are usually connected in parallel, which allows them to receive full supply voltage for operation. Switches used in electrical circuits to control these loads are connected in series, breaking the circuit if the switch opens.

You must understand the circuitry in air-conditioning, heating, and refrigeration control and power systems to do an effective job of installing and servicing the equipment.

We begin our study with a discussion of the basics of electric circuits.

FIGURE 2.1 A closed circuit   FIGURE 2.2 An open circuit

2.1 BASIC CONCEPTS OF ELECTRIC CIRCUITS

An **electric circuit** is the complete path of an electric current, along with any necessary elements, such as a power source and a load. When the circuit is complete so that the current can flow, it is termed **closed** or made (Figure 2.1). When the path of current flow is interrupted, the circuit is termed **open** or broken (Figure 2.2). The opening and closing of electrical switches connected in series with electrical loads control the operation of loads in the circuit.

All electric circuits must have a complete path for electrons to flow through, a source of electrons, and some electric device (load) that requires electric energy for its operation. Figure 2.3 shows a complete circuit with the basic components labeled. Electric circuits supply power and control loads through the use of switches.

An alternating current power supply is the most common source of the electric energy needed in the electric circuits of a heating, cooling, or refrigeration system. A direct current power supply, such as the dry cell battery, is often the source of electron flow in electric meters. Other than for meters, though, direct current is rarely used as a source of electric energy. Two special applications of direct current used today are electronic air cleaners and solid-state modules used for some types of special control, such as a defrost control or overcurrent protection.

The purpose of most electric circuits is supplying energy to a machine that does work. The most common device supplied with energy is an electric motor. Motors are used to rotate fans, compressors, pumps, and other mechanical devices that require a rotating motion. Automatic switches also require a power source to open and close them so they can start other electric devices. Electric circuits also supply energy to transformers, lights, and timers.

FIGURE 2.3 Basic electric circuit with circuit components labeled

FIGURE 2.4 Series circuit with four resistance heaters

There are several possible path arrangements that electrons may follow. The path arrangement is determined by the use or purpose of the circuit. The three types of circuits are series, parallel, and series-parallel. The series circuit allows only one path of electron flow. The parallel circuit has more than one path. The series-parallel circuit is a combination of the series and parallel circuits. In the following sections, we will look at each of these arrangements.

2.2 SERIES CIRCUITS

The simplest and easiest electric circuit to understand is the series circuit. The series circuit allows only one path of current flow through the circuit. In other words, the path of a series circuit must pass through each device in the entire circuit. That is, all devices are connected end-to-end within a series circuit. Figure 2.4 shows a series circuit with four resistance heaters.

Applications

Series circuits are incorporated in most control circuits used in heating, cooling, and refrigeration equipment. A **control circuit** is an electric circuit that controls some major load in the system. If all control components are connected in the circuit in series, the opening of any switch or component will open the circuit and stop the electric load, as shown in the circuit in Figure 2.5. (The symbols L_1 and L_2 in Figure 2.5 and in other figures represent the source of the voltage, the power supply. We will discuss circuit notation in more detail in Chapter 4.)

Series circuits are used in the electric circuitry of heating, cooling, and refrigeration equipment to operate the equipment and maintain a desired

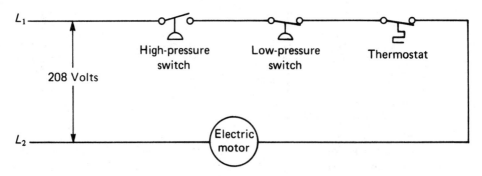

FIGURE 2.5 Series circuit with three switches for controlling an electric motor

temperature. Any electric switch or control that is placed in series with a load will operate that particular load. Figure 2.6 shows one circuit of a control system. The controls are connected in series with the device that is being controlled, in this case an electric motor.

The series circuit also contains any safety devices that are needed to maintain safe operation of the equipment components. Figure 2.7 shows a series circuit that is basically made up of safety devices designed to stop the compressor if an unsafe operating condition occurs. If any of the safety controls opens, the circuit will open and stop the compressor.

Calculations for Current, Resistance, and Voltage

The current draw in a series circuit is the same throughout the entire circuit because there is only one path for the current to follow. The current in a series circuit is shown by the following equation:

$$I_t = I_1 = I_2 = I_3 = I_4 = \cdots$$

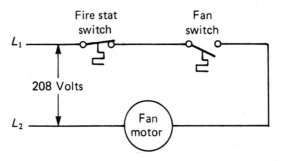

FIGURE 2.6 Control (series) circuit of a fan motor in a gas furnace. Both temperature controls must be closed for the fan to operate.

FIGURE 2.7 Control (series) circuit showing safety devices used on modern air-conditioning systems

(The centered dots $\cdots$ indicate that the equation continues in the same manner until all the elements of that particular circuit have been accounted for.)

The total resistance R_t in a series circuit is the sum of all the resistances in the circuit. The resistance of a series circuit is shown by the following equation:

$$R_t = R_1 + R_2 + R_3 + R_4 + \cdots$$

The voltage in a series circuit is completely used by all the loads in the circuits. The loads of the series circuit must share the voltage that is being delivered to the circuit. Thus, the voltage being delivered to the circuit will be split by the loads in the circuit.

The voltage of a series circuit changes through each load. This change is called the **voltage drop**. The voltage drop is the amount of voltage lost through any load or conductor. The voltage drop of any part of a series circuit is proportional to the resistance in that part of the circuit. The sum of the voltage drops of a series circuit is equal to the voltage being applied to the circuit. This is shown by the following equation:

$$E_t = E_1 + E_2 + E_3 + E_4 + \cdots$$

Ohm's law can be used for the calculations on any part of a series circuit or on the total circuit. Figure 2.8 shows a series circuit with four resistance heaters of different ohm ratings. The calculations for the total resistance, the amperage, and the voltage drop across each heater will be calculated using the circuit shown in Figure 2.8.

The total resistance can be calculated by adding the ohm rating of each heater.

Step 1: Use the formula $R_t = R_1 + R_2 + R_3 + R_4$.

Step 2: Substitute the values given in the figure into the formula:

$$R_t = 4 \text{ ohms} + 10 \text{ ohms} + 12 \text{ ohms} + 14 \text{ ohms}$$

FIGURE 2.8 Series circuit containing four resistance heaters with different ohmic values

Step 3: Solve the formula: $R_t = 40$ ohms.

We use Ohm's law to calculate the amperage draw (current) of the circuit.

Step 1: Use the formula

$$I = \frac{E}{R}$$

Step 2: Substitute the given values into the formula:

$$I = \frac{120}{40}$$

Step 3: Solve the formula: $I = 3$ amperes.

Now we use Ohm's law to calculate the voltage drop across each heater.

Step 1: Use the formula $E = IR$ for each resistance.

Step 2: Substitute the given values into the formula: $E_{d1} = 3 \times 4$.
(The symbol E_{d1} means the voltage drop across resistance 1.)

Step 3: Solve the formula: $E_{d1} = 12$ volts.

Step 4: Solve for each resistance using the same procedures that were used in Steps 2 and 3:

$$E_{d2} = 3 \times 10 = 30 \text{ volts}$$
$$E_{d3} = 3 \times 12 = 36 \text{ volts}$$
$$E_{d4} = 3 \times 14 = 42 \text{ volts}$$

Note that the total voltage E_t supplied to the circuit is equal to the sum of the voltage drops.

2.3 PARALLEL CIRCUITS

The parallel circuit has more than one path for the electron flow. That is, in a parallel circuit the electrons can follow two or more paths at the same

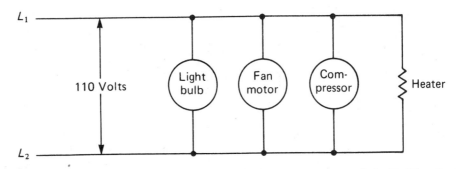

FIGURE 2.9 Parallel circuit with four components; each component is supplied with 110 volts of line voltage.

time. Electric devices (loads) are arranged in the circuit so that each is connected to both supply voltage conductors.

Parallel circuits are common in the industry because most loads used operate from line voltage. Line voltage is the voltage supplied to the equipment from the main power source of a structure and typically has a value of 110 volts or 230 volts. The parallel circuit allows the line voltage to reach all the electric loads connected in parallel, as indicated in Figure 2.9. Note that each load in the circuit is supplied by the line voltage of 110 volts.

Applications

Parallel circuits are used in the industry to supply the correct line voltage to several different circuits in a control system. Figure 2.10 shows a control system with several circuits in parallel being fed line voltage. Many different paths for electron flow are in this parallel hookup. Each circuit that is connected from L_1 to L_2 on the wiring diagram is in parallel with all the others and is being supplied with the line voltage.

Parallel circuits are used in all power wiring that supplies the loads of heating, cooling, or refrigeration systems. The electric loads of a system must be connected to the power supply separately or in a parallel circuit to supply the load with the full line voltage.

Calculations for Current, Resistance, and Voltage

There will be few occasions when field personnel are required to make calculations for a parallel circuit. This is usually done by the designer of the equipment. However, field personnel should be familiar with the basic concepts and rules of parallel circuits.

The current draw in a parallel circuit is determined for each part of the circuit, depending on the resistance of that portion of the circuit. The total

COMP: Compressor
CFM: Condenser Fan Motor
CH: Crankcase Heater
CT: Cooling Thermostat
LPS: Low Pressure Switch

FIGURE 2.10 Control system with several circuits (parallel); each circuit is supplied by the line voltage.

current draw of the entire parallel circuit is the sum of the currents in the individual sections of the parallel circuit. The current in each individual circuit can be calculated by using Ohm's law when the resistance and voltage are known. The total ampere draw of a parallel circuit is given by the following equation:

$$I_t = I_1 + I_2 + I_3 + I_4 + \cdots$$

The resistance of a parallel circuit gets smaller as more resistances are added to the circuit. The total resistance of a parallel circuit cannot be obtained by taking the sum of all the resistances. It is calculated by the following formula if two resistances are used:

$$R_t = \frac{R_1 \times R_2}{R_1 + R_2}$$

If three or more resistances are located in the circuit, the reciprocal of the total resistance is the sum of the reciprocals of all the resistances (the reciprocal of a number is one divided by that number). The following formula is used to calculate the resistance of a parallel circuit with more than two resistances:

$$\frac{1}{R_t} = \frac{1}{R_1} + \frac{1}{R_2} + \frac{1}{R_3} + \frac{1}{R_4} + \cdots$$

The voltage drop in a parallel circuit is the line voltage being supplied to the load. In other words, in a parallel circuit each load uses the total voltage being supplied to the load. For example, if 110 volts are supplied to a load, it will use the total 110 volts. The voltage being applied to each of the four components in Figure 2.9 is the same and is given by the following equation:

$$E_t = E_1 = E_2 = E_3 = E_4$$

Ohm's law can be used to calculate voltage, amperage, or resistance if the other two values are known. You can use Ohm's law to determine almost any condition in a parallel circuit, but pay careful attention to the individual sections of each complete circuit.

Example 1   What is the total current draw of the parallel circuit shown in Figure 2.11?

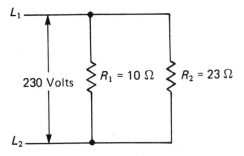

FIGURE 2.11 Parallel circuit for Example 1

Step 1:   First calculate the current draw for each individual circuit by using Ohm's law in the form

$$I = \frac{E}{R}$$

Step 2:   For I_1 substitute the given values for E and R_1 in the formula and solve:

$$I_1 = \frac{230}{10} = 23 \text{ amperes}$$

Step 3:   For I_2 substitute the given values of that circuit in the formula and solve:

$$I_2 = \frac{230}{23} = 10 \text{ amperes}$$

Step 4:   Use the formula $I_t = I_1 + I_2$.

Step 5: Substitute in the formula for I_t and solve:

$$I_t = I_1 + I_2$$

$$= 23 \text{ amperes} + 10 \text{ amperes} = 33 \text{ amperes}$$

Example 2 Find the total resistance of the parallel circuit in Figure 2.11.

Step 1: Use the formula

$$R_t = \frac{R_1 \times R_2}{R_1 + R_2}$$

Step 2: Substitute the known values in the formula:

$$R_t = \frac{10 \times 23}{10 + 23} = 6.97 \text{ ohms}$$

Example 3 What is the resistance of a parallel circuit with resistances of 3 ohms, 6 ohms, and 12 ohms?

Step 1: Use the formula

$$\frac{1}{R_t} = \frac{1}{R_1} + \frac{1}{R_2} + \frac{1}{R_3}$$

Step 2: Substitute the known values in the formula:

$$\frac{1}{R_t} = \frac{1}{3} + \frac{1}{6} + \frac{1}{12}$$

Step 3: Mathematical computation of this formula is sometimes difficult. If you have any problems, consult your instructor.

$$\frac{1}{R_t} = \frac{4}{12} + \frac{2}{12} + \frac{1}{12} = \frac{7}{12}$$

$$R_t = 1.71 \text{ ohms}$$

2.4 SERIES-PARALLEL CIRCUITS

The series-parallel circuit is seldom used in individual circuits in the industry. It is more often seen on the full wiring layout of an air-conditioning, heating, or refrigeration unit. This type of electric circuit is a combination of the series and parallel circuits, as shown in Figure 2.12. The series-parallel circuit is sometimes easier to understand when it contains only a few components. It becomes harder to understand when there are a large number of components.

Legend

LPS: Low-Pressure Switch
HPS: High-Pressure Switch
THER: Thermostat
C: Contactor
COMP: Compressor
CFM1: Condenser Fan Motor 1
CFM2: Condenser Fan Motor 2

FIGURE 2.12 Series-parallel circuit with four loads and controlling switches

The series-parallel circuit is often used to combine control circuits with circuits that supply power to loads, as shown in Figure 2.13. The circuit arrangement of most series-parallel circuits is designed so that all loads receive the correct voltage to operate switches and contacts in series with these loads thus controlling the operation of the system.

Any calculation of the values in a series-parallel circuit must be performed carefully because each portion of the circuit must be identified as series or parallel. Once the type of circuit has been determined, the calculations are made accordingly. The most common use of a series-parallel circuit in the industry is in the overall schematic diagram of a piece of equipment.

Applications

Series-parallel circuits are used in most types of heating, cooling, and refrigeration equipment in the industry. This circuit allows the required voltage to be supplied to electrical loads in the system. In the parallel circuits, the switches used to control the loads are connected in series. Refer again to Figure 2.13 that shows this type of circuit.

In a series-parallel circuit, both the series and parallel circuits are used to supply electrical power to the loads in the electrical systems. In Figure 2.13 the compressor, condenser fan motor, and indoor fan motor are connected

Legend

C:	Contactor
COMP:	Compressor
CFM:	Condenser Fan Motor
IFM:	Indoor Fan Motor
IFR:	Indoor Fan Relay
HP:	High-Pressure Switch
LP:	Low-Pressure Switch

Thermostat (R, G, and Y are
markings on thermostat terminals)

FIGURE 2.13 Wiring diagram of a packaged air-conditioning unit including the series-parallel circuit arrangement

in parallel to the supply voltage L_1 and L_2. The indoor fan relay coil and the contactor coil are in parallel with the transformer supplying voltage to the loads. Using parallel circuits enables the correct voltage to be supplied to the loads in the electrical circuitry of the equipment. Series circuits are used in electrical circuitry of air-conditioning equipment for control and safety switches that stop and start loads to maintain the correct temperature and safe operating conditions. Any switch connected in series with the loads will stop and start that load depending on the switches position. In Figure 2.13 the high and low pressure switches and thermostat are connected in series and control the contactor coil.

SUMMARY

The three types of electric circuits used in the industry are series, parallel, and series-parallel. The series circuit has only one path for electron flow. The most common type of electric circuit used in the industry is the parallel, which has more than one path for electron flow. This type of electric circuit allows line voltage to reach all electric loads. The series-parallel circuit is a combination of the series and parallel circuits.

The series circuit makes up most of the control circuits used in the industry because if any switch in the circuit opens, the load in the circuit will stop or start depending on the position of the switches. For example, if any switch that is connected in series with a load is open, the load will be deenergized; if all the switches in the circuit are closed, the load will operate. If a switch in the circuit used for temperature control is open, the load will be off. If the switch is closed or calling, the load will operate. Series circuits are used to control loads in the electrical system.

Parallel circuits are used in the industry and in equipment to ensure that proper voltage is supplied to the load. Parallel circuits are designed with more than one path for electron flow. In the electrical circuitry for equipment, the parallel circuit is used to supply the correct voltage to the electrical loads in the system. Parallel circuits are used in the wiring of houses to ensure that all receptacles are supplied with 110 volts.

The series-parallel circuit is a combination of series and parallel circuits. In the proper combination of these circuits, control systems supply voltage and control these circuits. Most electric control systems used to operate equipment the series-parallel circuit will be used.

The series and parallel circuits have different relationships among the voltage, amperage, and resistance. In a series circuit, the voltage is split among the electrical loads, which is the reason this circuit is not used to supply power to loads in equipment. The current in the series is equal in all parts of the circuit. The sum of all resistances in series is the total circuit resistance. In a parallel circuit, the voltage in all circuit parts is equal, thus is the reason they are used. The sum of each current flow in the circuit is the total current. The reciprocal of the total resistance is the sum of the reciprocals of all resistances.

QUESTIONS

1. What is a series circuit?

2. What is a control circuit?

3. How are series circuits used in air-conditioning equipment?

4. Why are series circuits used for most control circuits?

5. Why does the voltage drop in a series circuit?

6. What is a parallel circuit?

7. Why are parallel circuits used to feed electrical energy to loads?

8. Why are parallel circuits used in air-conditioning circuitry?

9. What is a series-parallel circuit?

10. Why are series-parallel circuits important in the circuitry used in air-conditioning equipment?

11. How would switches used as safety devices be connected with respect to the load they are protecting?

12. What happens if two 110-volt loads are connected in series?

13. Would it be possible to connect two safety controls in parallel with a load and properly protect the load?

14. Compare the calculations for voltage, amperage, and resistance in a series and a parallel circuit.

15. What are the requirements for an electrical circuit?

16. What is the resistance of a parallel circuit with resistances of 2 ohms, 4 ohms, 6 ohms, and 10 ohms?

17. What is the resistance of a parallel circuit with resistances of 10 ohms and 20 ohms?

18. What is the total amperage draw of a parallel circuit with ampere readings of 2 A, 7 A, and 12 A?

19. What is the voltage of a series circuit with four voltage drops of 30 volts each?

20. Draw a series-parallel circuit with one switch controlling one load.

LAB SECTION REFERENCE

For experiments and activities dealing with material covered in this chapter, refer to Lab Number 2, Series and Parallel Circuits.

3

Electric Meters

INTRODUCTION

Electricity is used in the control and operation of all refrigeration, heating, and air-conditioning systems. Approximately 80% of all service calls made by service technicians are diagnosed as electrical problems. Electrical meters, in most cases, are required for the proper diagnosis of these electrical problems. The large number of electrical problems encountered in the industry requires that industry personnel be able to correctly read and use electric meters. In the industry there are many types of electric meters. Some meters are for specific purposes while others are for the general day-to-day tasks that industry personnel perform.

The installation mechanic must be able to read electric meters to properly complete the initial start up and testing of equipment. The installation mechanic must be able to read and use electric meters to check the electrical characteristics of a newly installed refrigeration, heating, and air-conditioning system. No installation is complete without electrical checks of the system.

The service technician diagnoses and repairs problems found in refrigeration, heating, and air-conditioning systems. The service technician must be able to read and use all types of electric meters to quickly and efficiently locate electrical problems. Without electric meters and the knowledge to correctly use them, the service technician would face an almost impossible task in troubleshooting electrical problems.

An electric meter is a device used to measure the electrical characteristics of an electrical circuit. Electric meters are available in many different types and designs and must be selected for their use in the industry. Most meters are used by installation and service technicians. Meters must be durable and maintain adequate accuracy.

The electrical characteristics in a circuit that are most important to industry personnel are volts, amperes, and ohms. Although there are other important characteristics, these three are the most important. Most popular electrical meters are built around these three electrical characteristics. Most electrical meters used in the industry are capable of reading more than one electrical characteristic. The most common electrical meters used are the volt-ohm-milliammeter and the clamp-on ammeter with the ability to read volts and ohms.

The installation and service mechanic should be careful in selecting an electrical meter. Electrical meters should be selected for their everyday use in the industry. Technicians should select meters with the best ranges for the electrical characteristics that will be measured the most.

We will look at the basic principles used to measure electrical characteristics along with the various types and designs of electrical meters.

3.1 ELECTRIC METERS

An electric meter is a device used to measure some electrical characteristic of a circuit. The most common types of electric meters are the voltmeter, the ammeter, and the ohmmeter.

Basic Principles

Most electric measuring instruments make use of the magnetic effect of electric current. When electrons flow through a conductor in an electric circuit, a magnetic field is created around that conductor, as shown in Figure 3.1. This magnetic field is used to move the needle of a meter a certain distance, which represents the amount of the characteristic (volts, ohms, or

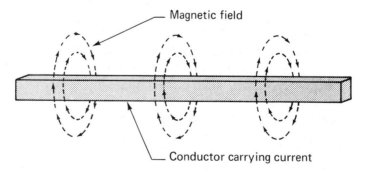

FIGURE 3.1 Magnetic field is produced around a conductor when current is flowing through the conductor.

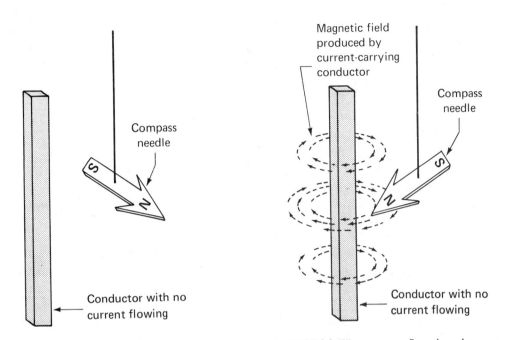

FIGURE 3.2 When there is no current flow, the compass reacts only to the magnetic field of the earth.

FIGURE 3.3 When a current flows through a conductor, the compass needle swings in line with the conductor's magnetic field.

amperes) being measured. The stronger the magnetic field, the larger the movement of the needle. The weaker the magnetic field, the smaller the movement of the needle.

If a compass is suspended next to a conductor that is not carrying an electron flow, the compass reacts only with the magnetic field of the earth and there is no other movement, as shown in Figure 3.2. However, when electrons flow through that same conductor, the compass needle swings in line with the conductor's magnetic field, as shown in Figure 3.3. The mechanical movement of the needle is caused by the magnetic field produced by the electron flow through the conductor. The larger the current flow, the stronger the magnetic field produced and the greater the needle movement on the scale. This simple principle is the basis of the meter movement in most electric meters.

Differences Among Meters

The differences among the various electric meters are not in the meter movements, except for digital meters, but in the internal circuits of the meters and in how the magnetic fields are created. For electrons to flow in an electric circuit, an electrical load must be present. This electron flow is somewhat different for the clamp-on ammeter, voltmeter, and ohmmeter. The clamp-on meter picks up the magnetic field through a set of laminated jaws on the meter. The voltmeter uses a resistor as a load to produce a magnetic field when voltage is applied to the circuit in the meter. The ohmmeter has its own power supply and uses the device being checked as the load to produce a magnetic field. All three meters use the same meter movement, unless the movement is digital. Their methods of loading and their power supplies are varied to attain the needle movement for reading the magnetic field.

Meters may be made in a combination and mounted in one case, or they may be completely separate. Figure 3.4 shows a clamp-on ammeter and a volt-ohm meter. The volt-ohm meter, as its name suggests, is used to measure either voltage or resistance, depending on the scale selected.

The meter movement created by the magnetic field around a conductor is reflected by the movement of the needle on a scale of an electric meter. This scale is usually broken down into several basic scales that have different ranges for voltage, amperage, and resistance. Some electric meters have a selector switch: the meter movement is shown on a certain point of the scale but the scale to be used must be determined by the person using the meter.

Digital meters are becoming more popular in the industry due to the reduction in cost over the past several years and the ease in reading. Another reason for the increasing popularity of the digital meter is the increased use of electronic control, which often times requires an electrical meter capa-

FIGURE 3.4 (A) Clamp-on ammeter with volt-ohm functions (B) Volt-ohm meter

FIGURE 3.5 Analog meter scale with needle

ble of accurately reading small voltages because of the lower signal voltages. An analog meter with its needle movement on a scale is shown in Figure 3.5. The mechanic must estimate the values between the lines on the meter scale. Digital meters can be read more accurately because there is no estimation as shown from the digital reading in Figure 3.6. The ease of reading a digital meter is certainly an advantage, and many technicians are using digital meters in the industry today.

Most digital meters utilize a 3½ or 4½ digit display. Figure 3.7 shows a 3½ digit display, and Figure 3.8 shows a 4½ digit display. The digit display can be explained by relating the ½ digit as being blank (0) or one that is the left digit of the meter reading. The 3 or 4 digits represent the next digits of the meter reading. For example, a 3½ digit reading will display a 4-digit reading with the left digit blank (0) or 1, such as 1999. This resolution determines the basic accuracy of the meter. Typical accuracy for a digital volt-ohm-milliameter is plus or minus 1% of actual reading compared with the typical accuracy of an analog meter of plus or minus 2% of full scale.

While analog meters depend on a magnetic field to move the needle, the digital meter makes use of Ohm's law to measure and display the electrical characteristics of the circuit.

Digital meters use two methods to protect the meter circuitry. Some digital meters are protected by internal circuits that detect an overload condition and then return the meter to normal operation. This function generally protects the ohm function from voltage overloads. Other digital meters are

FIGURE 3.6 Digital reading

FIGURE 3.7 3½ digit digital display

FIGURE 3.8 4½ digit digital display

FIGURE 3.9 Digital meters *(Courtesy of Amprobe Instruments, Lynbrook, NY) (Courtesy of A.W. Sperry Instrument Co., Inc., Hauppaugue, NY) (Courtesy of John Fluke Mfg., Co., Inc., Everett, WA)*

protected by fuses that have to be changed for the meter to again function. Several digital meters are shown in Figure 3.9.

3.2 AMMETERS

The ammeter uses the basic meter movement discussed in Section 3.1. The strength of the magnetic field determines the distance that the needle of the meter moves. Figure 3.10(a) shows the magnetic field and the meter movement when there is a high current flow through the conductor. Figure 3.10(b) shows the magnetic field and the meter movement when there is a low current flow in the conductor. The larger the current flow, the stronger the magnetic field grows and the greater the needle movement on the scale.

The ammeter measures current flow in an electric circuit. There are basically two types of ammeters used in the industry today: the clamp-on ammeter and the in-line ammeter. The clamp-on ammeter is the most popular because it is the easiest to use. You simply clamp the jaws of the meter around the conductor feeding power to the load that is producing the current draw. The clamp-on ammeter is shown in Figure 3.11. Figure 3.12 shows the new digital-readout clamp-on ammeter.

The in-line ammeter is used mostly on electric testing panels as a built-in

FIGURE 3.10 Scale of an ammeter showing current flow

component. It is also made with specially prepared leads to enable the service technician to use it as a plug-in unit. An in-line ammeter is shown in Figure 3.13. The in-line ammeter must be connected in series with the load that is producing the current draw to give the correct reading. In most cases, it is not practical to break a conductor and insert the in-line ammeter to

FIGURE 3.11 Clamp-on ammeters with volt-ohm functions

FIGURE 3.12 Digital ammeter   FIGURE 3.13 In-line ammeter

obtain an amperage reading. Therefore, the clamp-on ammeter is usually used.

The clamp-on ammeter is easy to use. Just follow a few simple rules. The jaws of the clamp-on ammeter are clamped around the conductor that is supplying a load or circuit. The magnetic field created by the current flowing through the wire is picked up by the jaws of the ammeter and funneled into the internal connection of the meter.

Never clamp the jaws of the meter around two wires to obtain an ampere reading. If the current flows in the wires are opposite, as they often are, the meter will read zero because the current flows cancel each other out. If the current flows are not opposite, the meter will read the current draw in both conductors. In either case, you will obtain an incorrect reading, which could cause you to incorrectly diagnose the problem.

When the ampere draw is small, you may have difficulty obtaining a true reading because of the small needle movement. The problem of small needle movement can be remedied by coiling the wire around the jaws of the meter. This allows the meter to pick up a larger current flow than is actually there. The meter will be more accurate because the current reading will fall in the midrange of the scale and can be easily read.

To obtain the correct ampere reading when this method is used, divide the ampere draw read by the number of loops going through the jaws of the meter. Figure 3.14 shows an ammeter with the conductor looped through the jaws three times. The correct reading can be obtained by dividing the meter reading, which is 4, by the number of loops through the jaws, which is 3. Thus, the ampere draw of the load is 1.33 amperes. Remember: The meter reading should always be divided by the number of loops carried through the jaws of the meter.

The clamp-on ammeter is one of the most valuable tools that industry personnel can carry. When you are installing a system, knowing the ampere draw of the equipment tells you if the unit is operating properly. You can also detect many electric circuit problems with a clamp-on ammeter. The ammeter can be purchased with other meters built into it. For example, a single clamp-on ammeter can be purchased that will read amperage, voltage, and resistance.

3.3 VOLTMETERS

The voltmeter is used to measure the amount of electromotive force available to a circuit or load. This is an important factor to heating, cooling, and refrigeration personnel because a wide range of voltages is presently used in this country.

Voltmeters range from simple to complex instruments containing many scales. The simplest voltmeter available is a small, inexpensive one capable

FIGURE 3.14 Taking a low-ampere reading with an ammeter by looping the wire through the jaws

FIGURE 3.15 Simple voltmeter

of distinguishing only between 110 and 230 volts (Figure 3.15). Several manufacturers build simple voltmeters that can read only voltages, but these are becoming increasingly difficult to obtain. More common is the volt-ohm meter, which reads both voltage and resistance. These are available in many forms, and service personnel should follow the instructions for the particular model being used. The common volt-ohm meter has three voltage scales and several voltage ranges. Some meters also have a high-voltage jack. Figure 3.16 shows three common types of volt-ohm meters. Figure 3.17 shows one of the new digital-readout volt-ohm meters available.

The voltmeter is designed much like the ammeter, but a resistor is added to the circuit to prevent a direct short and allow electrons to flow in the meter. The voltmeter uses two leads that are connected to jacks that lead to the internal wiring. To obtain a reading, the two leads must touch or be connected to the conductors supplying the load or to the circuit that transfers the electromotive force to the meter. The electrons flow through the leads into the meter through a resistor with a known ohm rating. The greater the voltage carried into the meter, the greater the magnetic field and the greater the needle movement.

When you do not know the voltage available to the equipment being worked on, you should start with the highest scale on the meter. Then change the meter setting until the needle falls in the midrange of the voltage scale. Never abuse a voltmeter by attempting to read a higher voltage than the range of the meter.

The voltmeter is almost a necessity for field personnel who have anything to do with the electrical section of equipment or with the installation or servicing of equipment. No heating, cooling, or refrigeration equipment should operate at an unsafe voltage, that is, voltage that is either too low or too

FIGURE 3.16 Three volt-ohm meters commonly used in the industry

FIGURE 3.17 Digital volt-ohm meter

high. All equipment is designed to operate at a voltage of 10 percent above or below the rating of the equipment. But in some cases the voltage may actually be more than the allowable figures. So field personnel should always check the supply voltage. Installation mechanics have not completed their installation unless they have checked the voltage available to the equipment. Service mechanics are often required to check the voltages to equipment as part of their troubleshooting job. In addition, the voltmeter can be used by service mechanics as a tool for diagnosing problems in the system. Thus, the voltmeter is a must for proper installation and service.

3.4 OHMMETERS

The ohmmeter is used to determine the operating condition of a component or a circuit. The ohmmeter can be used to find an open circuit, an open component, or a direct short in a circuit or component. It can also be used to measure the actual resistance of a circuit or a component.

The word **continuity** is used many times when referring to the use of ohmmeters. Continuity means that a particular circuit or component has a complete path for current to follow. An open component or circuit means that there is no resistance or infinite resistance in the circuit. The term **measurable resistance** means the actual resistance that is measured with the

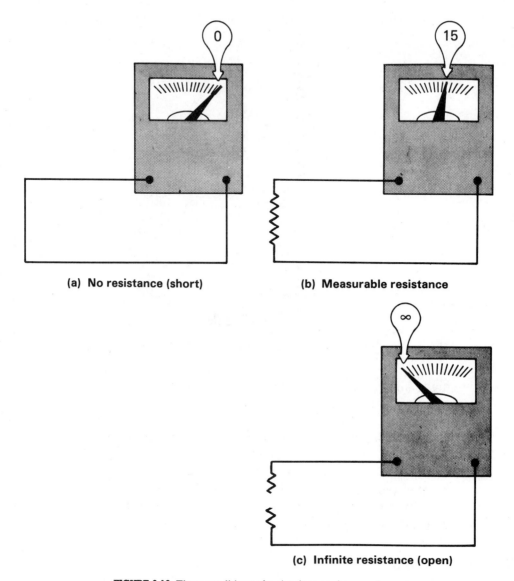

(a) No resistance (short)

(b) Measurable resistance

(c) Infinite resistance (open)

FIGURE 3.18 Three conditions of a circuit as read on an ohmmeter

ohmmeter. Figure 3.18 shows the three conditions as they might appear on the scale of an ohmmeter.

The ohmmeter is a valuable tool for diagnosing and correcting problems in electric circuits. In the industry many electric devices and circuits must be checked. The ohmmeter provides an easy method for checking circuits for opens (i.e., open circuits) and shorts (i.e., short circuits) and for measuring resistance.

An open circuit causes no noticeable needle movement in an ohmmeter because there is not a complete circuit. For example, an open (circuit) could occur in a blown fuse, a motor winding (the internal portion of a motor), and any condition where the electric circuit does not have a complete path for electrons to follow.

A direct short in an electric device or circuit causes problems because it means that two legs of the electric power wiring are touching, which causes an overload. In many cases a direct short means the wiring of the component is connected in some fashion. A closed switch is considered to be a short, but without this type of short, no heating, cooling, or refrigeration system would operate properly.

In many cases you will have to measure the resistance of a component to ensure that the component is in good operating condition. Most manufacturers make available to service personnel the exact ohmic value of motor windings and other components in the system.

The meter movement of an ohmmeter is designed and built for a very low current that is available from its own power source, usually a battery. Figure 3.19 shows the internal wiring of an ohmmeter. The ohmmeter works much like the ammeter and the voltmeter except for the small current that is supplied from the internal power source. The ohmmeter also uses a magnetic field to move the needle, but the magnetic field is created by a self-contained power source in the meter. The two leads of the ohmmeter are connected to the internal circuit of the meter, which contains a resistance and the power source. The amount of current the small battery can push

FIGURE 3.19 The internal wiring of an ohmmeter

through the device being tested indicates the resistance of that device and determines the needle movement.

Due to the low current that an ohmmeter is built to carry, it should never be connected to a circuit or device that is being operated. The function of the ohmmeter is merely to read the resistance of a device or circuit. Fortunately, most ohmmeters are equipped with some type of overload protection to protect the internal circuits of the meter if they are subject to line voltage.

Many types and designs of ohmmeters are available. In many cases the ohmmeter and voltmeter are combined in a dual-purpose meter. Some manufacturers build and market combinations of voltmeters, ohmmeters, and ammeters that are inexpensive. On this three-purpose meter, the ohmmeter is a low-range ohmmeter and cannot be used for many of the jobs that a service mechanic must do. The more expensive volt-ohm meters are more accurate and cover all ohm ranges. These meters will usually have at least three ohm scales (usually $R \times 1$, $R \times 100$, and $R \times 10,000$). Some have more ranges. Those additional ranges are useful for some troubleshooting operations. Refer again to Figure 3.16 that shows three different volt-ohm meters used today.

SUMMARY

Electricity plays an important part in the industry. Most equipment has some type of electric control system, even if it is powered by some other means. Thus, it is important for industry personnel to be familiar with the basic types of electric meters. About 80% of the problems with equipment or systems are electrical, which shows that electric meters are important.

The ammeter is used to measure the current flow in an electric circuit. Two types of ammeters are used in the industry today. The clamp-on ammeter is the most frequently used. With this type of ammeter, it is only necessary to clamp the jaws of the meter around the conductor feeding the circuit or load and to read the amperage. The in-line ammeter must be put in series with the load or circuit to read the amperage. Because of the time required to do this, the in-line ammeter is seldom used.

The voltmeter is used to measure the voltage of an electric circuit. The voltmeter will have to be connected in parallel to the circuit to determine the voltage supplying the circuit. Voltage measurements can determine the source voltage, voltage drop, and a voltage imbalance. The first step in troubleshooting an air-conditioning system is to determine if voltage is available to the equipment or system.

The ohmmeter is used to measure the resistance of a circuit or device in ohms. The ohmmeter must be used with the circuit power off to prevent damage to the meter. Most ohmmeters have a power source built in the meter. The ohmmeter is used to determine the condition of electrical devices used in air conditioning such as motors, heaters, contactor coils, solenoid valves, and other components. A continuity check is another use of the ohmmeter, which determines if a complete path is available for current to flow.

The decrease in cost of digital meters has made them more popular in the industry. Digital meters can be used without any interpolation of the reading, making them an advantage over analog meters where interpolation must be made with reference to the needle on the scale. Mechanics should choose the meter that best suits their needs in the industry. With the proper care, meters will give mechanics many years of good service.

QUESTIONS

1. What are the three most common electric meters used in the industry?

2. Why should industry personnel be competent at using electric meters?

3. Almost all analog meters make use of the _____.

4. How does an ammeter work?

5. What are the two types of ammeters? Which type is more commonly used in the industry?

6. What is the result on the meter reading of clamping the jaws of a clamp-on ammeter around two wires?

7. How can very small ampere draws be measured with a clamp-on ammeter?

8. How does an analog voltmeter work?

9. Air-conditioning or refrigeration equipment can operate properly at __ percent above or below its rated voltage.

10. If a service mechanic has no idea of the voltage available to a unit, what procedure should be followed on reading the voltage?

11. What precaution should be taken when using an ohmmeter?

12. What does the term continuity mean?

13. What is a short circuit?

14. What is an open circuit?

15. What factor should be considered when purchasing an electric meter?

16. What is the reaction of an ohmmeter to a measurable resistance, a short circuit, and an open circuit?

17. An in-line ammeter must be connected to a circuit in _____.

18. A voltmeter is connected to a circuit in _____.

19. State the differences in internal circuitry among the ohmmeter, voltmeter, and ammeter.

20. What is the difference between an analog meter movement and a digital meter movement?

21. Give three advantages of a digital electric meter.

22. What basic concept is used in a digital meter to calculate the circuit characteristics being measured?

23. What is the basic accuracy of most analog and digital meters?

24. What is the difference between a 3½ and 4½ digit resolution on a digital meter?

25. How are digital meters protected?

LAB SECTION REFERENCE

For experiments and activities dealing with material covered in this chapter, refer to Lab Number 1, Reading Electrical Meters.

Components, Symbols, and Circuitry of Air-Conditioning Wiring Diagrams

INTRODUCTION

Because of the complexity of today's air-conditioning, heating, and refrigeration systems, industry personnel should be able to read and interpret all kinds of wiring diagrams. Electric wiring diagrams contain a wealth of information about the electrical installation and operation of the equipment. The installation mechanic depends on the wiring diagram for the correct installation of the wiring to the unit. The service mechanic uses the electrical diagrams as a guide in troubleshooting the electric system of a unit.

It would be impossible for wiring diagrams to be composed of photographs of various components of the equipment. They would be too large and in many cases too complex due to the number of wires that are carried to certain devices. Thus, symbols are used in wiring diagrams to represent such system components as compressors, indoor fan motors, thermostats, pressure switches, and heaters. Industry personnel must be able to identify most symbols and know where to look up the remainder.

Most manufacturers use similar symbols for each type of electric component, although there are some minor differences in symbols between some major manufacturers. Thus, a knowledge of the basic symbols is essential if you are to be successful in the industry.

We begin our study with a discussion of the various types of electric loads found in the industry and the basic symbol used for each device.

4.1 LOADS

Loads are electric devices that consume electricity to do useful work. Loads are devices such as motors (Figure 4.1), solenoids (Figure 4.2), resistance heaters (Figure 4.3), and other current-consuming devices. The size of loads

FIGURE 4.1 Electric motor

FIGURE 4.2 A solenoid used to operate a contactor

FIGURE 4.3 A resistance heater

varies from devices with a small current draw, such as a light bulb, a small fan motor, and solenoids, to large motors that could use upwards of 100 amperes.

Loads are the most important part of a heating, cooling, or refrigeration system because they do all the work in the system. Loads operate **compressors**, which compress and transfer refrigerant in a system. They operate fans, which move air. They operate the solenoid part of a relay, which starts and stops loads. Also, loads operate other devices that perform useful work. Industry personnel should be able to recognize the common symbols for loads and know where to look up the symbols for little-used loads, because each electric wiring diagram is composed of symbols and their interconnecting wires.

In the following paragraphs, we will take a close look at several different kinds of loads used in the industry.

Motors

A **motor** is an electric device that consumes electric energy to rotate a device in an electric system. Motors are used in the industry to rotate devices such as a compressor (Figure 4.4), a condenser fan motor (Figure 4.5), pumps (Figure 4.6), and other units that require rotating movement. Motors are the largest and most important loads in heating, cooling, and refrigeration systems.

FIGURE 4.4 Compressor

FIGURE 4.5 A condenser fan motor on a small air conditioner

FIGURE 4.6 A pump used as a condenser water pump

(a) Most commonly used symbol

(b) Alternate symbol

FIGURE 4.7 Symbols for an electric motor

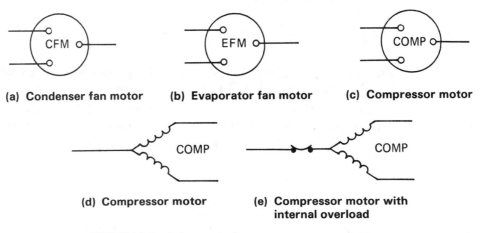

(a) Condenser fan motor   **(b) Evaporator fan motor**   **(c) Compressor motor**

(d) Compressor motor   **(e) Compressor motor with internal overload**

FIGURE 4.8 Symbols representing some common uses of motors

The symbols shown in Figure 4.7 are the most common symbols used to represent motors.

A letter designation tells you what purpose the motor serves in the system. Figure 4.8 shows several symbolic representations of different uses of motors. Careful attention should be given to symbols representing motors because in some cases a motor has an internal overload, as shown in Figure 4.8(e).

Solenoids

The **solenoid** is a device that creates a magnetic field when energized and causes some action to an electric component such as a relay or valve. Two common solenoids used to operate a relay and a contactor are shown in

FIGURE 4.9 Two solenoids used to operate a relay

FIGURE 4.10 (a) Solenoid valve and solenoid coil (b) Symbol

Figure 4.9. The solenoid is considered a load because it consumes electricity to do useful work.

Solenoids are devices that open and close to control some element in a system. Solenoid valves are valves that open and close, stopping or starting a flow. Solenoid coils used in relays and contactors will be discussed later in this chapter. Some common solenoid valves are hot-gas solenoids, reversing-valve solenoids, and liquid-line solenoids. Figure 4.10(a) shows a solenoid valve and a solenoid coil, and Figure 4.10(b) shows its symbol.

Heaters

Heaters are loads that are found in many systems and wiring diagrams. A heater takes electric energy and converts it to heat. In some cases, electric resistance heaters are used to heat homes. Heaters might also be used to heat a small object or area. The symbol for all heaters is the same. Only a letter designation tells you specifically why the heater is used. Figure 4.11 shows the symbols used for heaters, along with some common letter designations.

(a) **Heater**

CH
(b) **Crankcase heater**

SUPP H
(c) **Supplementary heater**

FIGURE 4.11 Symbols for common uses of resistance heaters

(a) Red   **(b) Green**   **(c) Blue**

FIGURE 4.12 Symbols for signal lights showing the color of the light

Signal Lights

A **signal light** is a light that is illuminated to denote a certain condition in a system. The letter inside the signal light symbol denotes the color of the signal light, as shown in Figure 4.12. Signal lights come in a variety of colors and are not limited to the colors shown in Figure 4.12. A signal light is used to show that a piece of equipment is operating or that it is operating in an unsafe condition. Signal lights are usually energized when a piece of equipment or component is started.

4.2 CONTACTORS AND RELAYS

Contactors and **relays** are devices that open and close a set or sets of electric contacts by the action of a solenoid coil. The contactor or relay is composed of a solenoid and the contacts. A relay is shown in Figure 4.13. A

FIGURE 4.13 Relay   FIGURE 4.14 Contactor

contactor is shown in Figure 4.14. When the solenoid is energized, the contacts will open or close, depending on their original position (that is, if they were open, they will close, and vice versa).

In an air-conditioning control system, we must have some method of controlling loads. In most cases, a relay or contactor is used. Relays and contactors are widely used in control systems. Thus, it is essential that industry personnel be able to identify the symbols for relays and contactors.

The only difference between a relay and contactor is the size of the device. A contactor is simply a large relay. Usually, the devices are distinguished by their rated current flow. A contactor can carry 20 amperes or more. A relay is designed to carry less than 20 amperes. It is likely for contactors to be used where the ampere draw of a device is more than 20 amperes. However, a relay would rarely be used to carry over 20 amperes.

Contactors and relays play an important part in the control system of any air conditioner, refrigerator, or heater. For example, contactors and relays are used to stop and start different loads in a refrigeration system. Compressors, in most air-conditioning systems, are controlled by a contactor or magnetic starter. Relays can be used for pilot duty, that is, for controlling another relay or contactor. The most important fact to remember is that most control systems have many relays and at least one contactor. These relays or contactors always control some load.

Relays and contactors are composed of two parts: the contact and the coil, or solenoid. The contact makes the electrical connections. Figure 4.15 shows the symbol for a **pole**, or contact, of a relay or contactor. The term "pole" refers to one set of contacts. However, in some cases the relay or contact might have two or three poles, which means two or three sets of contacts. The coil or solenoid, the second part of the relay, is **energized** (voltage is supplied) and, through a magnetic field, closes the contact or contacts. Either symbol shown in Figure 4.16 can be used to represent a relay or contactor coil. The symbol for the relay or contactor is the same if each has the same number of poles and if their purpose is basically the same, with the exception of the ampere rating of the device.

All symbols are usually shown in the **deenergized** position. This means that there is no electric potential to the coil of the device. Figure 4.15 shows a "normally open" pole in the deenergized position.

The term **normally** refers to the position of a set of contacts when the

FIGURE 4.15 Symbol for a normally open pole of a relay or contactor

FIGURE 4.16 Symbols for relay or contactor coil; either symbol may be used for each device.

FIGURE 4.17 Normally closed set of contacts

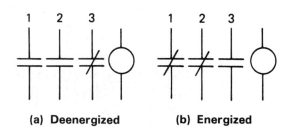

(a) Deenergized   **(b) Energized**

FIGURE 4.18 Symbols showing relays deenergized and energized

device is deenergized. The contacts in Figure 4.15 are called **normally open**. This means they would close when the relay or contactor is energized. Shown in Figure 4.17 is a relay with a set of **normally closed** contacts. These contacts would open if the relay was energized.

The terms "normally open," "normally closed," "energized," and "deenergized" are important in understanding relays and contactors on wiring diagrams. Figure 4.18(a) shows a relay with two normally open contacts and one normally closed contact in the deenergized position (with no voltage to the coil). Figure 4.18(b) shows the same contacts in the energized position (with voltage to the coil). In the deenergized position, the current will not flow through contacts 1 and 2, but current will flow through contact 3. In the energized position, the current flow is through 1 and 2 but not through 3.

4.3 MAGNETIC STARTERS

A **magnetic starter** is the same type of device as a contactor in terms of the ampere rating of the device. But the magnetic starter has a means of overload protection in it, whereas the contactor has none. Figure 4.19 shows a picture of the magnetic starter and its symbol. The principle of operation of the magnetic starter will be covered in Chapter 9.

4.4 SWITCHES

An electric **switch** is a device that opens and closes to control some load in an electric circuit. Electric switches can be opened and closed by temperature, pressure, humidity, flow, or by some manual means. You must become

(a) Symbol   (b) Magnetic starter

FIGURE 4.19 Magnetic starter

familiar with the symbols used for switches because in most cases they control the loads in the system. The symbol will also indicate what is initiating the action of the switch.

A manually operated switch is a switch that is opened and closed by manual force. Figure 4.20 shows a simple manually operated switch. The poles of a manual switch are the number of contacts that are included in the switch. The **throw** indicates how the switch may be operated. For example, a single-pole–single-throw switch has one set of contacts and two positions: an open and a closed position, as shown in Figure 4.20. A double-pole–double-throw switch has two sets of contacts and two positions, as shown in Figure

FIGURE 4.20 Single-pole–single-throw manual switch

FIGURE 4.21 Double-pole–double-throw manual switch

4.21. Symbols for these two switches and for two other basic types of manual switches are shown in Figure 4.22.

There are other types of manual switches used in the industry. The **disconnect switch** is used to open and close the main power source to a piece of equipment or load. Figure 4.23 shows a three-pole disconnect switch and its symbol. The **push-button switch**, as shown in Figure 4.24, is a switch used to open and close a set of contacts by pressing a button. The symbols for the normally closed and the normally open push-button switches are also shown in Figure 4.24.

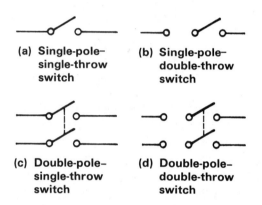

(a) Single-pole–
single-throw
switch

(b) Single-pole–
double-throw
switch

(c) Double-pole–
single-throw
switch

(d) Double-pole–
double-throw
switch

FIGURE 4.22 Symbols for manual switches

(a) Switch

(b) Symbol

FIGURE 4.23 Three-pole fusible disconnect switch

(a) Switch

(b) Symbol for normally closed push-button switch

(c) Symbol for normally open push-button switch

FIGURE 4.24 Push-button switch

The most important type of switch in a control system is the mechanically operated switch. **Thermostats** are mechanically operated switches used in most control systems. Thermostats are said to be mechanically operated because the temperature-sensing element moves a set of contacts by a mechanical linkage. Thermostats are designed for heating, cooling, or both. The cooling thermostat is designed to close on a temperature rise and open on a temperature fall. The heating thermostat is designed to open on a temperature rise and close on a temperature fall. The symbols for these two types of thermostats, shown in Figure 4.25, indicate their function. Figure 4.26 shows a modern thermostat.

Pressure switches are used for different functions in modern control circuits. The purpose of the pressure switch determines whether it opens or closes on a rise or fall in pressure. The pressure range of the switch is not part of the symbolic representation. Figure 4.27 shows the symbols for pressure switches. Letter designations in the symbols often denote the pressure ranges and purposes of the switches. Figure 4.28 shows some common pressure switches used in the industry.

(a) Heating thermostat; opens on temperature rise

(b) Cooling thermostat; closes on temperature rise

FIGURE 4.25 Symbols for heating and cooling thermostats

FIGURE 4.26 Thermostat

(a) Opens on rise in pressure

(b) Closes on rise in pressure

FIGURE 4.27 Symbols for pressure switches

FIGURE 4.28 Some common pressure switches

4.5 SAFETY DEVICES

Safety devices are important in today's modern systems. Components are becoming more expensive each year. Thus, it is vital that these components be protected from adverse conditions such as low voltage, high ampere draw, and overheating. It is for this reason that you should become familiar with symbols for safety devices. Overloads and safety devices are sometimes a combination of a load and a switch. They differ from the relay in their purpose and overall design.

All motors are designed to operate on a certain current draw. If for some reason this rating is exceeded, the motor must be cut off immediately to prevent damage and possible destruction of the component. A burned out motor is often caused by a malfunction in the safety devices.

The **fuse** is the simplest type of overload device. The fuse is effective against a large overload, but it is less effective against small overloads. The fuse is nothing more than a piece of metal designed to carry a certain load. Any higher load will cause the fuse to break the circuit. Figure 4.29 shows two symbols for a fuse. Figure 4.30 shows some common fuses in use today.

The second type of overload device is designed to protect the motor against small and large overloads. They are divided into two categories: thermal and magnetic. The **thermal overload** is operated by heat, and the **magnetic overload** is operated by magnetism, which is directly proportional to the current draw.

The thermal overload can be a **pilot duty** device, which breaks the control circuit and locks the motor out. The pilot duty types of overloads are most common on motors larger than 3 horsepower. Or the thermal overload can be a line voltage device, which breaks the power line to the component being protected.

FIGURE 4.29 Symbols for fuses

FIGURE 4.30 Some common fuses

The bimetal element is the simplest of the thermal overloads. When it gets warm, it warps to open the circuit, as shown symbolically in Figure 4.31. Some bimetal elements are furnished with heaters, as shown symbolically in Figure 4.32. The heater allows the bimetal disc to react to an overload quicker because the current flow is proportional to heat.

The thermal overload relay, whose symbols are shown in Figure 4.33, is a simple device with a thermal element and a switch that opens on a rise in temperature.

The magnetic overload symbol is the same as the symbol for a relay with one normally closed contact. The current flow is relayed to the overload coil. Since current flow is proportional to the strength of the magnetic field, the relay can be designed to energize only on a high current draw. Figure 4.34 shows a magnetic overload and its symbol. The letter designa-

(a) Closed   **(b) Open**

FIGURE 4.31 Symbols for bimetal overload

Heaters

FIGURE 4.32 Symbol for three-wire bimetal overload

FIGURE 4.33 Symbols for a thermal overload relay

(a) Symbols   **(b) Magnetic overload device**

FIGURE 4.34 Magnetic overload

tion of this device will distinguish between the magnetic overload and the common relay.

4.6 TRANSFORMERS

The **transformer** decreases or increases the incoming voltage to a desired voltage. In most air-conditioning control circuits, it is not practical to pull large wires for a long distance. Therefore, a 24-volt control circuit, which is safer, less expensive, and a better method of control, is used. Figure 4.35 shows a transformer and its symbol. The voltage is also given with the symbol in some cases.

4.7 SCHEMATIC DIAGRAMS

Most modern heating, cooling, and refrigeration systems are becoming more complex with more controls and safety devices. Advances in controls and control systems require you to be able to read **schematic diagrams**. If you are able to read schematic diagrams, you will know what the unit should be doing.

The schematic diagram is the most useful and easiest to follow of any electric diagram. The schematic diagram tells how, when, and why a system works as it does. In most cases, service technicians use schematic diagrams to troubleshoot control systems. The schematic wiring diagram includes the

(a) Symbol   **(b) Transformer**

FIGURE 4.35 Transformer

symbols and the line representations so the user can easily identify loads and switches along with the circuits.

All electric circuits contain a source of electrons, a device that uses electron flow, and a path for the electrons to follow. In most cases, the source of electrons is an alternating current voltage supply. The device using the electron flow is a motor, heater, relay coil, or any other load device. The path for the electrons to follow is a wire or any type of conductor.

In the schematic diagram, the source of electrons, the power supply, is represented by two lines drawn downward and listed as L_1 and L_2, as shown in Figure 4.36. There is a potential difference of 230 volts between L_1 and L_2. If a path is created between L_1 and L_2, current will flow.

All electrical loads in the unit are placed between L_1 and L_2, along with the switches controlling the load. Figure 4.37 shows a complete circuit in schematic form with a compressor and the switch (thermostat) that controls it. When the switch is closed, the compressor will run. In Figure 4.37, the source of electrons is from L_1 and L_2, the path is the connecting wire, and the device using the electron flow is the compressor. The compressor operates when the thermostat is closed.

Figure 4.38 shows a full schematic diagram similar to the diagrams you will be using on the job. All schematic diagrams are broken down into a circuit-by-circuit arrangement. Most schematic diagrams contain a legend that cross-references the components and their letter designation to the name of the component. Look for the legend in Figure 4.38.

4.8 READING SIMPLE SCHEMATIC DIAGRAMS

Schematic diagrams are used throughout the industry to show a systematic layout of a control system. Schematic diagrams range from simple to com-

FIGURE 4.36 Schematic diagram of a complete circuit

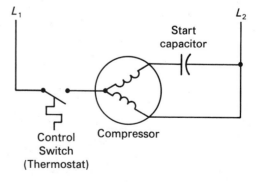

FIGURE 4.37 Schematic diagram of a complete circuit with a control

FIGURE 4.38 Complete schematic diagram for small packaged unit *(Courtesy of Westinghouse Electric Corp., Central Residential Air Conditioning Division)*

plex, depending on the use. Most residential systems use a simple schematic with few components. A large commercial and industrial system usually has a complex diagram, which is harder to read and understand. However, the layout arrangement of a schematic diagram is the same whether the system is residential or commercial and industrial. This section will cover two basic schematics, circuit by circuit.

Schematic Wiring Diagram of a Simple Air Conditioner

The first schematic we will discuss is shown in Figure 4.39. The source of electrons comes from L_1 and L_2. The potential difference of the circuit in Figure 4.39 is 230 volts. The diagram represents a simple air conditioner with line voltage controls. The indoor fan motor is controlled by a switch and must be on for the compressor circuit to operate.

Figure 4.40 shows the basic circuit of the indoor fan motor (IFM) operated with a double-pole–single-throw switch. Note that one pole is not used in the IFM circuit. When the switch is closed, the IFM will operate.

Figure 4.41 shows the IFM circuit and the contactor circuit. The contactor coil circuit has two switches that control the circuit. The purpose of the manual double-pole–single-throw switch used in the contactor coil circuit is to keep the contactor coil from being energized except when the IFM is operating. The thermostat energizes the contactor coil to maintain the thermostat setting. When the contactor coil is energized, the C_1 and C_2 contacts of Figure 4.39 close and start the compressor. If the thermostat becomes satisfied, the contactor is deenergized and the compressor is stopped.

FIGURE 4.39 Schematic wiring diagram of a simple air conditioner

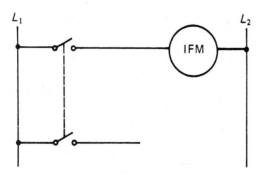

FIGURE 4.40 Indoor fan motor circuit of the schematic diagram

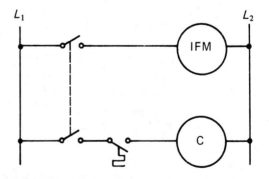

FIGURE 4.41 Indoor fan motor circuit with contactor coil circuit

Schematic Diagram of a Packaged Air-Conditioning Unit

The next schematic we will discuss is that of a simple air-cooled package unit with a low-voltage control system. The complete schematic is shown in Figure 4.42. A packaged unit is a unit that is built in one piece.

Most residential air-conditioning systems have a 24-volt control circuit (again see Figure 4.42) to operate the electric control components. When a 24-volt control system is incorporated, more components are required because a transformer must be used to reduce line voltage to the 24-volt low-voltage supply.

The schematic diagram of Figure 4.42 is divided into two sections. The portion above the transformer is the line voltage section. The portion below

Legend

C: Contactor
COMP: Compressor
CFM: Condenser Fan Motor
IFM: Indoor Fan Motor
IFR: Indoor Fan Relay
HP: High-Pressure Switch
LP: Low-Pressure Switch

Thermostat (R, G, and Y are markings on thermostat terminals)

FIGURE 4.42 Schematic diagram of a packaged air-conditioning unit

the transformer is the 24-volt control circuit. The only connection between the two circuits is the transformer and the fact that the contacts of the low-voltage controlled relays are connected into the line portion of the control system.

The schematic for the low-voltage thermostat used on the unit is shown in Figure 4.43. The thermostat is a combination of several switches along with the actual temperature-detecting portion of the control, which in this case is a cooling thermostat. The switch in the upper left of the thermostat diagram is a system switch. It cuts the total control system off; that is, it stops the entire system from operating. The fan switch allows the fan to be operated alone on either the "on" or the "manual" setting. When the fan switch is set on automatic, the fan operates only when the compressor operates. compressor operates.

The low-voltage control portion of the system opens and closes relays and contactors in the line voltage portion of the circuit. The low-voltage control circuit also operates the compressor and condenser fan motor. The contactor coil of the low-voltage system (Figure 4.42) is energized through the cooling thermostat, assuming that the system switch is in the "on" position. When the contactor coil is energized, its contacts close, starting the compressor and condenser fan motor. The high-pressure and low-pressure switches in the contactor coil circuit will deenergize the contactor coil whenever the refrigeration system reaches an unsafe operating condition, be it high or low pressure in the refrigeration system.

The upper portion of the schematic diagram in Figure 4.42 is divided into three separate circuits. These circuits are discussed individually in the following paragraphs.

Figure 4.44 shows the compressor and condenser fan motor circuit. The

FIGURE 4.43 Schematic diagram of the 24-volt thermostat with all switches

FIGURE 4.44 Schematic diagram of the compressor and condenser fan motor circuit

circuit contains the compressor and condenser fan motor. The compressor and condenser fan motor are stopped and started by the contacts of the contactor.

The indoor fan motor circuit is shown in Figure 4.45. The indoor fan motor starts when the contacts on the indoor fan relay close.

Figure 4.46 shows the indoor fan relay coil circuit and the low-voltage circuits of the indoor fan motor. The fan is operated by closing the indoor fan relay contacts. When the thermostat fan switch is in the "on" position, the indoor fan relay energizes and closes the IFR contacts, operating the indoor fan continuously until the fan switch is moved to the automatic position. When the fan switch is in the automatic position, the indoor fan relay coil is energized whenever the thermostat is closed.

The following steps give the sequence of operation of the air-conditioning unit. Refer again to Figure 4.42 as you read each step.

1. The contactor coil is energized by the cooling thermostat when the system switch is in the "on" position. When the contactor closes, the compressor and condenser fan motor start.
2. The fan motor operates continually with the fan switch in the "on" position because the IFR coil is energized, closing the contacts. If the fan switch on the thermostat is in the automatic position, the fan operates when the compressor is running.
3. The high-pressure and low-pressure switches are safety devices that deenergize the contactor if either of the switches opens.

The schematic diagrams for residential and small commercial systems are

FIGURE 4.45 Schematic diagram of the indoor fan motor circuit

FIGURE 4.46 Schematic diagram of the indoor fan motor and low-voltage circuits usually simple and easy to understand. It is important for service personnel to be able to interpret wiring diagrams so they can service the equipment. A schematic wiring diagram can tell a service mechanic how and when a piece of equipment operates as it does.

4.9   READING ADVANCED SCHEMATIC DIAGRAMS

Schematic Diagram of a Heat Pump

The last two schematic wiring diagrams discussed have been primarily air-conditioning systems. The heat pump is an air-conditioning system that simply reverses the refrigeration cycle to provide heating. This brief explanation of a heat pump should help you understand the following detailed explanation of the heat pump schematic.

The refrigeration cycle of a heat pump is basically the same as any air-conditioning system except for the reversing valve, which reverses the flow of refrigerant in the system. When a heat pump is operating in the air-conditioning mode, the indoor coil is the evaporator and the outdoor coil is the condenser. In the heating mode, the cycle reverses; the outdoor coil becomes the evaporator, and the indoor coil becomes the condenser. The reversing valve accomplishes the flow reversal in the refrigeration cycle. The heat pump can effectively cool in the summer and heat in the winter by reversing the flow of refrigerant. The heat pump removes the heat from the inside of a structure in the summer and expels the heat to the outside air, but in the winter the heat is absorbed from the outside air and rejected in the structure.

When the heat pump is operating in the heating mode, there must be a way to defrost the outside coil. This is done by reversing the cycle and using the cooling mode of operation to defrost the outside coil. In the defrost mode, a set of resistance heaters is energized to keep the indoor fan from moving cold air and causing drafts in the conditioned area. Many methods are used in the industry to defrost a heat pump.

The efficiency of the heat pump decreases as the outdoor temperature drops, and at a certain point, called the *balance point*, the heat pump can no longer supply sufficient heat to properly condition the conditioned area. Supplementary electric resistance heaters are added to the circuit to provide the additional heat needed to properly condition the space.

A two-stage heating–one-stage cooling or a two-stage heating–two-stage cooling thermostat is used to operate most heat pumps, depending on the control circuitry designed by the manufacturer. The heat pump diagram that will be covered uses a two-stage heating–two-stage cooling thermostat.

FIGURE 4.47 Typical complete operation schematic of heat pump

The heat pump that will be discussed is a split-system heat pump, which means the unit is divided into an outdoor section and an indoor section. The thermostat is mounted in the conditioned space.

The heat pump schematic will be covered in four stages: (1) cooling cycle, (2) heating cycle, (3) defrost cycle, and (4) resistance heating cycle. The explanation of the schematics will be presented in a step-by-step operating procedure.

A heat pump schematic is shown in Figure 4.47 with all the electrical components required for the complete heat pump system with reference numbers that can be used with the explanation. The legend used for the heat pump schematic is shown in Figure 4.48 with the reference letter or number in parentheses. The legend will be used for all four explanations.

Legend

AH:	Supplementary Heat Contactor	(19)
BH:	Supplementary Heat Contactor	(18)
CA:	Cooling Anticipator	
CR:	Run Capacitor	(4 & 6)
CPR:	Compressor	
D:	Defrost Relay	(9)
DFT:	Defrost Timer	(7)
DT:	Defrost Termination Thermostat	(8)
F:	Indoor Fan Relay	(15)
FM:	Manual Fan Switch	(F)
HA:	Heating Anticipator	
HTR:	Heater	
IOL:	Internal Overload Protection	
LT:	Light	
LVTB:	Low-voltage Terminal Board	
MS:	Compressor Motor Contactor	(5, 3, & 12)
MTR:	Motor	
ODA:	Outdoor Temperature Anticipator	(G)
ODS:	Outdoor Temperature Sensor	(13)
ODT:	Outdoor Thermostat	
RHS:	Resistance Heat Switch	(C)
SC:	Switchover Valve Solenoid	(11)
SM:	System Switch	(A)
TNS:	Transformer	(16)
TSC:	Cooling Thermostat	(B & D)
TSH:	Heating Thermostat	(E & H)

FIGURE 4.48 Legend of diagrams in Figures 4.47–4.53

The Cooling Cycle. The cooling cycle schematic of the heat pump that will be explained is shown in Figure 4.49. Only the electrical components that are functional in the cooling mode of operation are shown on the diagram.

The following step-by-step sequence describes the cooling cycle operation shown in the heat pump in Figure 4.49.

Step 1: Disconnect 1 (1) and disconnect 2 (20) are in the "on" position with the thermostat system switch (A) in the "off" position. Disconnect 1 (1) supplies power to the outdoor unit, thus supplying power to the trickle circuit through the start winding of the compressor motor (5) and the run capacitor (4) acting as a crankcase heater for the compressor. Disconnect 2 (20) supplies power to the indoor unit, thus supplying power to the transformer (16), which provides 24 volts to the control circuitry. The outdoor thermostat sensor (13) is energized through the outdoor temperature anticipator (G) in the thermostat whenever the disconnect for the indoor unit is turned on.

Step 2: First Stage Cooling—When the thermostat system switch (A) is moved to the "auto" position, the cooling anticipator is energized and provides heat to the thermostat bimetal during the off cycle. Due to the voltage drop across the cooling anticipator, the motor starter (12) will not energize. The first stage cooling thermostat (B) closes and completes the 24-volt circuit from terminal R to terminal O of the thermostat, thus energizing the switchover valve solenoid coil (11).

Step 3: Second Stage Cooling—The second stage cooling thermostat (D) closes and completes the 24-volt circuit from R to Y, thus energizing the motor starter (12), and from R to G, thus energizing the fan relay (15) after a 0.7° to 1.5° F temperature rise above the temperature of the first stage cooling thermostat. Simultaneously, the contacts (3 and 5) of the motor starter close, energizing the compressor and the outdoor fan motor; the indoor fan motor is energized through the indoor fan relay contacts (19). Once the second stage cooling thermostat closes, the cooling anticipator is bypassed.

The Heating Cycle. The heating cycle schematic of the heat pump that will be explained is shown in Figure 4.50. Only the electrical components that are functional in the heating mode of operation are shown on the diagram.

The following step-by-step sequence describes the heating cycle operation shown in the heat pump in Figure 4.50.

FIGURE 4.49 Typical cooling cycle schematic of heat pump

FIGURE 4.50 Typical heating cycle schematic of heat pump

Step 1: Disconnect 1 (1) and disconnect 2 (20) are in the "on" position with the thermostat system switch (A) in the "off" position. Disconnect 1 (1) supplies power to the outdoor unit, thus supplying power to the trickle circuit through the start winding of the compressor motor (5) and the run capacitor (4) acting as a crankcase heater for the compressor. Disconnect 2 (20) supplies power to the transformer (16) and provides 24 volts to the control circuitry. The outdoor thermostat sensor (13) is energized through the outdoor temperature anticipator (G) in the thermostat whenever the disconnect for the indoor unit is turned on.

Step 2: First Stage Heating—When the thermostat system switch (A) is moved to the "auto" position, the first stage heating thermostat (E), when closed, completes the 24-volt circuit from terminal R to terminal Y, thus energizing the motor starter (12), and from terminal R to terminal G, thus energizing the indoor fan relay (15). The heating anticipator is energized when the first stage heating thermostat closes. Simultaneously, the contacts (3 and 5) of the motor starter close, energizing the compressor and the outdoor fan motor; the indoor fan motor is energized through the indoor fan relay contacts (19). The second set of contacts (15) of the indoor fan relay provides a fan interlock for the heater control circuit.

Step 3: Second Stage Heating—The second stage heating thermostat closes after a temperature drop of 0.7° to 1.5° F. When the second stage heating thermostat closes, the 24-volt circuit from terminal R to terminals W and U energizes the supplementary heat contact coil (19) and the blue light on the thermostat. If the outdoor thermostat (14) is closed, the W circuit can also energize supplementary heat contactor (18).

The Defrost Cycle. When operating in the heating mode at temperatures below 40° F outdoors, the heat pump coil will accumulate frost or ice. The defrost cycle schematic of the heat pump that will be explained is shown in Figure 4.51. Only the electrical components that are functional in the defrost mode of operation are shown on the diagram. Defrost is accomplished by putting the heat pump in the cooling mode, so the outdoor coil becomes the condenser long enough to defrost the coil.

The following step-by-step sequence describes the defrost cycle operation shown in the heat pump in Figure 4.51.

FIGURE 4.51 Typical defrost cycle schematic of heat pump

Step 1: The heat pump is operating in the heating mode.

Step 2: Initiation of the defrost cycle to remove frost or ice from the outdoor coil requires that two conditions be met: (1) The lower circuit of the outdoor coil must be below 26° F, as detected by the defrost termination thermostat (8). (2) The defrost timer (7 and 8) must have completed the time requirement of 45 or 90 minutes compressor run time. If these two conditions are met, the timer cam switch (7) will close and complete a circuit through the closed defrost termination switch (8) and the defrost time switch (8) to the defrost relay coil (9). When the defrost relay is energized, defrost relay contacts 1 (11) and 3 (11) complete the circuitry to the supplementary heater contactor (18) and the switchover valve solenoid coil (11). Contact 2 of the defrost relay opens the circuit to the outdoor fan motor (10) and provides a holding circuit for the defrost relay coil.

Step 3: Termination of the defrost cycle is normally accomplished by the defrost termination thermostat (8) opening when the liquid temperature leaving the outdoor coil reaches approximately 52° F. If the defrost termination thermostat does not terminate the defrost cycle within 10 minutes, the override switch in the defrost timer will open and deenergize the defrost relay and thus terminate the defrost cycle.

The Resistance Heating Cycle. Most heat pumps are equipped with an emergency heat mode that allows the consumer to isolate the pump part of the heat mode and heat the residence by supplementary electric resistance heat. The emergency heat mode allows the consumer to maintain comfort of the conditioned space even though the normal heating mode is inoperative because of mechanical failure. The emergency heat cycle of the heat pump that will be explained is shown in Figure 4.52. Only the electrical components that are functional in the emergency heat mode of operation are shown on the diagram.

The following step-by-step sequence describes the emergency heat cycle operation shown in the heat pump in Figure 4.52.

Step 1: The emergency heat switches 1 and 2 (C) are mechanically linked to each other within the thermostat. When emergency heat switch 1 is moved from the normal position to the emergency position, terminal Y is removed from the circuit, thus isolating the motor starter and making terminal X2, which energizes the supplementary heat contactor coils (18 and 19) if the outdoor thermostat

FIGURE 4.52 Typical resistance heat cycle schematic of heat pump

(14) is closed. If the outdoor thermostat is open, the supplementary heat relays will be energized by the second stage heating thermostat when needed.

The entire heat pump schematic is shown in Figure 4.53; it can be understood when the mechanic looks at each mode of operation separately.

4.10 PICTORIAL DIAGRAMS

The **pictorial diagram**, also referred to as a label or line diagram, is intended to show the actual internal wiring of the unit. The pictorial diagram shows all the components of the control panel as a blueprint, including all the interconnecting wiring. It does not show the unit to scale, however. Components that are not shown in the control panel itself are shown outside the panel and labeled. The pictorial diagram is used to locate specific components or wires when troubleshooting from a schematic diagram. A typical pictorial diagram used in the industry is shown in Figure 4.54.

It is difficult to determine how a system operates by a pictorial diagram, and only an experienced mechanic can follow a complex pictorial diagram. Thus, most air-conditioning personnel use the schematic diagram to find the cause of the problem. Then they use the pictorial diagram to locate the position of the component at fault. In cases where the wiring is simple, however, a pictorial diagram may be the only diagram furnished with the equipment.

The **factual diagram** consists of a pictorial diagram along with a schematic diagram. Many air-conditioner manufacturers supply factual diagrams so service personnel can locate the relay or component in the control panel.

4.11 INSTALLATION DIAGRAMS

The **installation diagram** is used to help the installation electrician to wire the unit properly. The diagram gives specific information about terminals, wire sizes, color coding, and breaker or fuse sizes. The diagram does not provide details about equipment operation because the electrician has no need for this information. Figure 4.55 shows an installation diagram. The installation wiring diagram shows little internal wiring and is therefore almost useless to industry personnel.

SUMMARY

Loads are devices that use electricity to do useful work. Figure 4.56 gives a review of the symbols used for solenoids, motors, and heaters, the typical

FIGURE 4.53 Typical complete operation schematic of heat pump

FIGURE 4.54 A typical pictorial diagram used in the industry *(Reproduced by permission of Carrier Corporation. Copyright © 1977 Carrier Corporation, Syracuse, NY)*

To power supply — 230 V (for-1A, -1C) 60 Hertz
To power supply — 230/208 V (for-2A, -2C) 60 Hertz
Single phase, 25 A Fusetron recommended not to
exceed 30 A

FIGURE 4.55 A typical installation diagram *(Courtesy of The Singer Company, Climate Control Division)*

(a) Motor

(b) Solenoid

(c) Heater

FIGURE 4.56 Review of symbols used for loads

FIGURE 4.57 Review of symbols for contactors and
relays

loads found in the industry. Most symbols have some type of letter designation to identify more clearly the component referred to.

Loads are controlled by relays and contactors, which share the same symbol and perform similar tasks. The major difference between relays and contactors is the amount of current each can carry. If a compressor is being operated by a device, you can assume the device is a contactor. If a small fan motor is being operated by a device, you can assume the device is a relay. A relay is used for small loads, and a contactor is used for large loads. Figure 4.57 reviews the symbols for some of these devices.

Relays and contactors are controlled by switches. Some of the switches used in the industry are manual, push button, thermostat, and pressure. Thermostats are made for two purposes: to operate a heating or a cooling system. The symbols for thermostats denote whether they are used for heating or cooling. Pressure switches are much the same as thermostats; their symbols also denote which way they open or close and under what condition. Pressure switches can be used for low or high pressure and are usually denoted by letter designations.

In any system using motors, protective devices are important to prevent damage to the motors or to larger components of the system. The most important type of safety devices is for motor protection. A fuse, magnetic overload, thermal overload line break, thermal overload pilot duty, or a thermal overload relay could be used. Many overloads are built directly into the larger components.

Transformers are devices that increase or decrease the incoming voltage to some desired voltage. Transformers are used in the industry mainly in control circuits.

Schematic diagrams tell air-conditioning, heating, or refrigeration personnel when and why a system works as it does. Schematic diagrams show the symbols for devices and the interconnecting wiring of a unit in a circuit-by-circuit arrangement. Schematic diagrams are used most frequently by service personnel to troubleshoot equipment and systems.

Pictorial diagrams show an exact layout of the control panel with the external components shown outside the panel and labeled. The pictorial diagram can be used as a troubleshooting diagram on a simple system, such as a window air conditioner. In most cases, pictorial diagrams are used to find the placement of a component in the panel. Factual diagrams are a combination of the schematic and pictorial, with each shown separately.

Installation diagrams are used to help the installation electrician correctly connect the wiring to the unit.

QUESTIONS

1. What are the three types of electrical diagrams used in the heating, cooling, and refrigeration industry?

2. What is a load?

3. What is the major load of an air-conditioning system?

4. Identify the following symbols for loads:

a.

b.

c.

d.

e.

5. What is the major difference between a contactor and a relay?

6. What do the terms "normally open" and "normally closed" refer to?

7. What is the difference between a magnetic starter and a contactor?

8. Identify the following symbols for relays and contactors:

a. b. c.

d. e.

9. Why are heating and cooling thermostats different?

10. Identify the following symbols for switches:

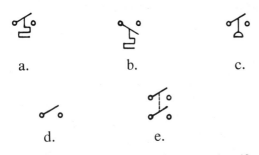

a. b. c.

d. e.

11. What is the purpose of an overload?

12. What is a magnetic overload?

13. Identify the following symbols for safety devices:

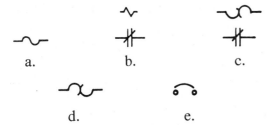

a. b. c.

d. e.

14. Draw the symbol for a transformer.

15. What are the requirements for an electric circuit?

16. What is a schematic diagram? What is its purpose?

17. What is a legend in a schematic diagram?

18. How is a pictorial diagram laid out?

19. What is a factual diagram?

20. What is the purpose of an installation diagram?

21. True or false. A solenoid is a device that opens or closes to control some element in a system.

22. Draw the symbol for a heater.

23. The purpose of the signal light in a system is to _____.

24. A contactor or relay is composed of _____.

25. True or false. Loads are controlled by switches, and switches are controlled by contactors or relays.

26. What do we mean when we say that a component is "energized"?

27. Name three basic switches used in the industry and draw their symbols.

28. What is the most important type of switch in a control system?

29. What is the purpose of a fuse in a system? What is the symbol for a fuse?

30. A bimetal element is _____.

31. True or false. A schematic diagram tells the service mechanic how to wire a system.

32. All schematic diagrams are broken down into _____.

Use Figure 4.53 to answer the following questions.

33. What is the purpose of the defrost cycle of a heat pump?

34. What two factors initiate the defrost mode in the diagram?

35. What type of thermostat is used with the heat pump in the diagram?

36. Explain the operation of the switchover valve and its purpose in the circuitry.

Use Figure 4.38 to answer the following questions.

37. The HP and LP switches are connected in _____ with the compressor motor.
    a. series
    b. parallel

38. The compressor motor is protected by an _____ overload.
    a. internal
    b. external

39. The condenser fan motor is protected by an _____ overload.
    a. internal
    b. external

40. The contactor is controlled by _____.
    a. the disconnect switch
    b. the high-pressure switch
    c. the cooling thermostat
    d. none of the above

41. With the fan switch in the "on" position, the indoor fan will _____.
    a. run when the cooling thermostat is closed only

b. run when temperature exceeds 95°F
    c. run continuously
    d. do none of the above

42. The primary voltage of the transformer is _____.
    a. 230 volt
    b. 208 volt
    c. 115 volt
    d. 208/230 volt

43. The compressor and condenser fan motors start _____.
    a. at separate times
    b. at the same time

44. The compressor and condenser fan motors are connected in _____.
    a. series
    b. parallel

Use Figure 12.11 to answer the following questions.

45. The control circuit of the unit is a __.
    a. line voltage circuit
    b. low-voltage circuit

46. The thermostat cycles _____.
    a. the compressor
    b. the indoor fan motor
    c. the compressor and indoor fan motor
    d. none of the above

47. The holding relay and time assembly work together as _____.
    a. an operating control
    b. an anti-short-cycling function
    c. an indoor fan delay function
    d. none of the above

48. The outdoor fan relay is energized by the thermostat starting the _____.
    a. compressor

b. indoor fan motor
c. outdoor fan motor
d. crankcase heater

49. The contactor coil is energized by the closing of _____.
   a. the thermostat
   b. the high-pressure switch
   c. the holding relay contacts (3 and 1)
   d. the timer contacts (A and A1)
   e. all of the above

50. If the unit is calling for cooling, the holding relay coil is energized by the

   _____.

   a. HR contacts (3 and 1)
   b. HR contacts (5 and 4)
   c. timer contacts (A and A1)
   d. OFR contacts

LAB SECTION REFERENCE

For experiments and activities dealing with material covered in this chapter, refer to Lab Number 3, Electrical Symbols Used in the HVAC Industry, and Lab Number 4, Reading Schematic Diagrams.

5

Alternating Current, Power Distribution, and Voltage Systems

INTRODUCTION

Two types of current are used in the heating, cooling, and refrigeration industry today: **direct current (DC)** and **alternating current (AC)**. Current is the flow of electrons. Direct current is an electron flow in only one direction. Alternating current is an alternating (back and forth) flow of electrons; that is, the electrons reverse their direction of flow at regular intervals. Direct current will not be discussed in this chapter because it is used in the industry only for special applications.

Most current produced by electric utilities is alternating current. In the rare instances that direct current is needed, it can be produced by a direct current generator or a rectifier. It is usually produced by the consumer. Direct current has limited use in the industry; it is used mostly for refrigeration transportation equipment, electronic air cleaners, and electronic control components.

Alternating current is used in most heating, cooling, and refrigeration equipment. Alternating current equipment is cheaper, more rugged, and more trouble free than direct current equipment. In addition, alternating current is easier to produce than direct current.

Because of the popularity and wide use of alternating current, it is important for industry personnel to be familiar with its theory. In addition, personnel should be familiar with the way power is distributed by electric utilities and the many types of voltage-current systems available.

We begin our study with a discussion of some basic ideas of alternating current.

5.1 BASIC CONCEPTS OF ALTERNATING CURRENT

Alternating current, or AC, is an electron flow that alternates, flowing in one direction and then in the opposite direction at regular intervals. It is produced by cutting a magnetic field with a conductor.

The **sine wave** is often used as a graphical representation of alternating current. Figure 5.1 shows a sine wave, the graph of alternating current. The letter X represents a conductor in a particular position as it is rotated in the magnetic field. At point A, X is at a potential of 0. However, the potential (voltage) increases when the conductor is rotated through the magnetic field until it reaches point B, where the potential peaks. When the conductor (X) is rotated from B to C, the potential decreases until at point C the potential is again 0. The direction of flow from point A to point C is termed positive. The direction is reversed on the lower half of the sine wave (point C back to point A) and is termed negative. From point C the potential increases until it reaches point D and peaks. The potential decreases from point D until the conductor reaches point A, where the potential is again 0. The curve shows the conductor from point A through a complete clockwise revolution back to point A.

While the conductor is rotating from point A to point C, one positive alternation occurs. That is, the curve starts from 0, reaches the peak at point B, and returns to 0 at point C. The next alternation is negative as X goes from point C back to point A in the same manner as the first alternation.

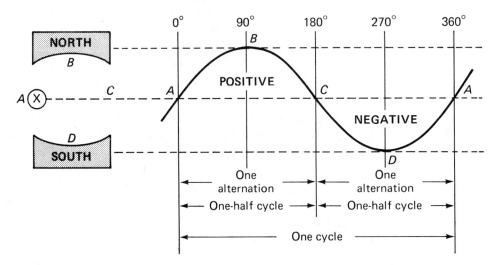

FIGURE 5.1 Sine wave of alternating current

Cycles and Frequency

When the conductor rotates through one complete revolution, it has generated two alternations, or flow reversals. Two alternations (changes in direction) equal one cycle (refer again to Figure 5.1). One cycle occurs when the **rotor**, or conductor, cuts the magnetic field of a north pole and a south pole.

The **frequency** of alternating current is the number of complete cycles that occur in a second. The frequency is known as **hertz** (Hz), but many times it is referred to as **cycles**. In most locations in the United States, the common frequency is 60 hertz. In some isolated cases in the United States and in some foreign countries, 25 hertz is the frequency used. The disadvantage to using 25 hertz is that the reversals can be detected by the human eye in light fixtures. The frequency of 60 hertz is considered standard frequency, but 50 hertz is also used in some cases.

Effective Voltage

Because alternating current starts at 0, reaches a peak, and then returns to 0, there is always a variation in voltage and an effective value has to be determined. Alternating current reaches a peak at 90 electrical degrees. This is known as **peak voltage**. The **effective voltage** of an alternating current circuit is 0.707 times its highest or peak voltage. This value is determined with respect to direct current so the effective voltage is equal to one direct current volt. Figure 5.2 shows a comparison of effective voltage versus peak voltage in an alternating current circuit. Alternating current amperage is 0.707 times the peak amperage. All electric meters are calibrated to read effective voltage and amperage.

FIGURE 5.2 Effective voltage versus peak voltage

Voltage-Current Systems

Alternating current is available to the consumer at several different voltages and with different current characteristics. The four basic voltage-current characteristics available are 230 volt-single phase-60 hertz, 230 volt-three phase-60 hertz, 208 volt-three phase-60 hertz, and 460 volt-three phase-60 hertz. These designations are often abbreviated; for example, 230 volt-single phase-60 hertz is abbreviated as 230 V-1ϕ-60 Hz. The V is the abbreviation for volts; ϕ (Greek letter phi) is the symbol for phase; and Hz is the abbreviation for hertz (which means the same as cycles).

Phase

The **phase** of an AC circuit is the number of currents alternating at different time intervals in the circuit. Single-phase current would have only a single current, while three-phase current would have three.

Alternator

Alternating current is produced by an **alternator**. The alternator is made up of a winding or set of windings called the stator and a rotating magnet called the **rotor**. The number of windings used depends on the desired phase characteristic of the current. When the rotor is rotated through the field of the winding, an alternating electrical current is produced.

Single-phase current is produced by an alternator with one winding in the stator as shown in Figure 5.3. Three-phase current is produced by an alter-

FIGURE 5.3 Winding arrangement of single-phase alternator

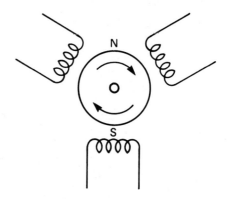

FIGURE 5.4 Winding arrangement of three-phase alternator

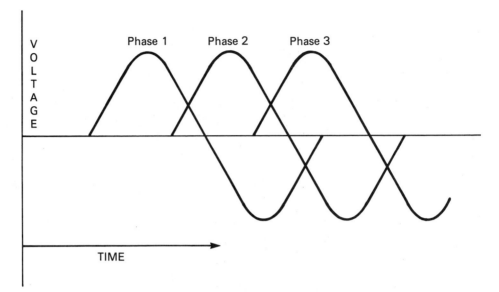

FIGURE 5.5 Sine wave of three-phase current

nator with three windings in the stator that are wound 120° apart as shown in Figure 5.4. The sine wave of three-phase current is shown in Figure 5.5.

Inductance and Reactance

It is a common assumption that when the voltage is strongest, the amperage is also strongest. However, this is not always the case. The amperage of a circuit determines the strength of the magnetic field in the circuit. If the current increases, so will the magnetism. If the current decreases, so will the magnetism. The fluctuation of the magnetic strengths in an AC circuit, and in conductors cutting through more than one magnetic field, *induces* (causes) a voltage that counteracts the original voltage. This effect is called **inductance**.

The inductance in an AC circuit—that is, the effect of the magnetic fields—produces an out-of-phase condition between the voltage and amperage. The induction of the original voltage produces a second voltage due to the magnetic fields collapsing. This effect causes the voltage to *lead* the amperage, as shown in Figure 5.6.

Resistance in a direct current circuit is the only factor that affects the current flow. AC circuits are affected by resistance, but they are also affected by **reactance**. Reactance is the resistance that alternating current encounters when it changes flow. There are two types of reactance in AC circuits: inductive reactance and capacitive reactance. **Inductive reactance** is the opposition to the change in flow of alternating current, which produces an out-

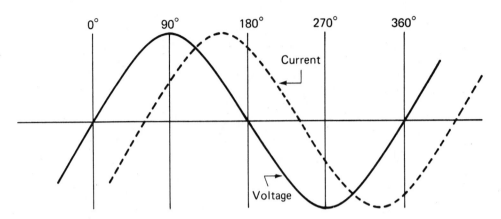

FIGURE 5.6 Sine wave of voltage and current when they are out of phase. Voltage leads the current in an AC circuit due to the effect of inductance.

FIGURE 5.7 Sine wave of voltage and current when they are out of phase. Current leads the voltage in an AC circuit due to capacitive reactance.

of-phase condition between voltage and amperage, as shown in Figure 5.6. **Capacitive reactance** is caused in AC circuits by using capacitors. When a capacitor is put in an AC circuit, it resists the change in voltage, causing the amperage to lead the voltage, as shown in Figure 5.7. The sum of the resistance and reactance in a AC circuit is called the **impedance**.

Power

As we have said, the voltage and current in an AC inductive circuit are out of phase with each other. When voltage and current are out of phase, they are not working together. Thus, the power (wattage) of the circuit must be calculated by using a voltmeter and an ammeter. This calculation gives the **apparent wattage**. A watt meter would measure the true power (wattage). The ratio between the true power and the apparent power is called the **power factor** and is usually expressed as a percentage. The following equation expresses this:

$$PF = \frac{\text{true power (measured)}}{\text{apparent power (calculated)}}$$

5.2 POWER DISTRIBUTION

Direct current was used in the beginning to supply consumers with their electrical needs, but it had many disadvantages. Transmission for a long distance without using generating stations to boost the direct current was impossible. The inability to raise and lower DC voltages and the necessity to use large transmission equipment were other problems. Alternating current can be transmitted with much less worry than direct current and has become the ideal power to supply to consumers because of its flexibility. Most equipment used in the industry incorporates alternating current as its main source of power.

Electric power is generated by rotating turbines through the use of gas, oil, coal, hydropower, or atomic energy. The rotating turbines have the effect of a rotating conductor in a magnetic field. The electricity is generated inside the plant and transferred outside the plant, where it is boosted to a large, easily transmitted voltage. This is frequently as high as 220,000 volts. The alternating current is then transmitted to a substation, where its voltage is reduced to around 4,800 volts with the use of a step-down transformer. From the substation, the power is distributed to transformers that step down the voltage to a usable voltage. Or the power is supplied directly to consumers who use their own transformers to reduce the voltage. Figure 5.8 shows a diagram of electric power transmission from the generating plant to the consumer.

In the following sections we will discuss the four common voltage systems available to consumers.

5.3 230 VOLT-SINGLE-PHASE-60 HERTZ SYSTEMS

Single-phase alternating current exists in most residences. Most small heating, cooling, and refrigeration equipment can be purchased for use on single phase. Any domestic appliance that operates on 110 volts is single-phase equipment. It is important to understand the makeup of the 230 V-1ϕ-60 Hz system because of its popularity in small commercial buildings and resi-

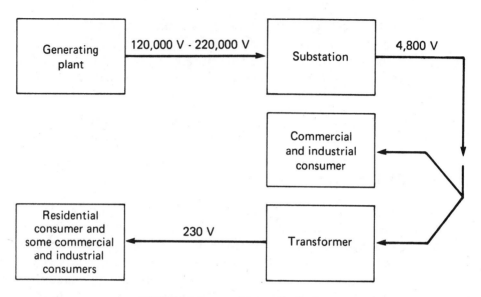

FIGURE 5.8 Layout of power distribution

dences. Air-conditioning personnel must be able to determine the type of voltage system available to ensure proper installation and operation of the equipment.

In some older structures it is still possible to find a single-phase-two-wire system. This voltage system was used when few electric appliances were available and there was no need for 230-volt systems. The system uses two wires, one as a hot wire and one as a neutral. **Hot wires** are conductors that actually supply voltage to an appliance. The neutral is a wire connected to the ground. In the voltage system shown in Figure 5.9, 110 volts is displaced through L_1 but must be connected to a neutral to make a complete circuit.

The most common voltage system found today is the 230 V-1ϕ-60 Hz system. This voltage system consists of three wires, two hot wires and one ground. A schematic for the 230-volt system is shown in Figure 5.10. The figure shows the system being used as a switch to break down the power being fed to a piece of equipment, such as a walk-in cooler, an electric furnace, or an air-conditioning unit. In this system, connecting either hot leg, L_1 or L_2, and the neutral directly to a load will supply 115 volts. Connecting L_1 and L_2 directly to a load will supply 230 volts.

The electric utility uses a transformer to produce the 230 V-1ϕ-60 Hz system. The transformer hookup of this system is shown in Figure 5.11. Pay close attention to the secondary side of the transformer. A transformer takes voltage at one value and by induction changes it to another value. Figure 5.11 shows a transformer with a 4800-volt primary winding and a 230-volt

FIGURE 5.9 Schematic for a 110 volt-single phase-60 hertz system using two wires

FIGURE 5.10 Schematic for a 230 volt-single phase-60 hertz system using three wires

FIGURE 5.11 Transformer hookup of a 230 V-1ϕ-60 Hz system

secondary winding. The primary winding, or input, is connected to the initial voltage, which is 4800 volts in Figure 5.11. The secondary winding produces the output, or new voltage, which is 230 volts in Figure 5.11.

All equipment can be operated at plus or minus 10% of the rated voltage. For example, a piece of equipment rated at 230 volts could operate on a minimum voltage of 207 volts and a maximum voltage of 253 volts. The electric utility maintains a plus or minus 10% of the correct supply voltage. Air-conditioning equipment has a tendency to operate more satisfactorily on a maximum voltage than on the minimum voltage.

5.4 THREE-PHASE VOLTAGE SYSTEMS

Three-phase alternating current is common in most commercial and industrial structures. Three-phase electrical services supply three hot legs of power with one ground to the distribution equipment and then on to the equipment. Three-phase power supplies are more versatile than single-phase supplies because of the different voltage systems that are available. Thus, it is essential to understand the uses and advantages of three-phase voltage systems.

Three-phase electric power supplies have many advantages over the single-phase power supply when used in commercial and industrial applications. In most cases, residences do not use enough electric energy to warrant a three-phase power supply. The electric power consumer can buy three-phase electric power cheaper than single-phase. Three-phase electric motors require no special starting apparatus, eliminating one trouble spot in the building and servicing of motors. Three-phase power offers better starting and running characteristics for motors than does single-phase. Many large electric motors used in the industry are available for three-phase only, preventing many structures from using single-phase power supplies. Thus, most commercial and industrial structures are supplied with three-phase power.

The only disadvantage to three-phase systems is the higher cost of electric panels and distribution equipment.

There are basically two types of three-phase voltage systems used in commercial and industrial wiring systems: the **delta transformer** hookup, which will supply 230 V-3ϕ-60 Hz, and the **Y transformer** hookup, which will supply 208 V-3ϕ-60 Hz or 460V-3ϕ-60 Hz.

5.5 230 VOLT-THREE-PHASE-60 HERTZ "DELTA" SYSTEM

The 230 V-3ϕ-60 Hz, or delta, system is used basically in structures that require a large supply to motors and other three-phase equipment. The delta system is usually supplied to a structure with four wires, which include three hot legs and a ground, but in some rare cases it is supplied with three hot wires only. Figure 5.12 shows the wiring layout of a 230 V-3ϕ-60 Hz system. This system is unique in that it contains a high leg, which in Figure 5.12 is L_1. Connecting L_1 to neutral, or ground, provides a range of 180 volts to 208 volts. Connecting L_1 to L_2, L_2 to L_3, or L_1 to L_3 provides 230 volts. Connecting L_2 to G or L_3 to G provides 120 volts. If one hot leg of a three-phase delta system is lost, only single-phase voltage is supplied and a single-phase condition exists. This condition can easily damage any three-phase equipment.

The transformer secondary hookup of a three-phase delta system is shown in Figure 5.13. The delta system takes its name from the Greek letter delta Δ, which resembles the shape of the hookup, as can be seen in Figure 5.13. The high leg of the voltage is due to the fact that the transformer winding

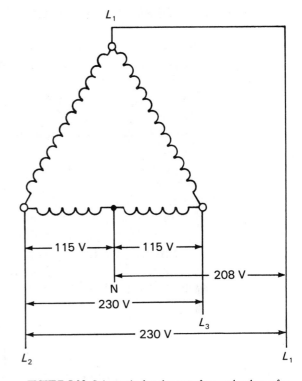

FIGURE 5.12 Schematic for a 230 volt-three phase-60 hertz system using four wires

FIGURE 5.13 Schematic for the transformer hookup of a 230 V-3ϕ-60 Hz system showing the delta transformer secondary hookup

between L_1 and G is longer than the windings between L_2 and G and between L_3 and G.

The 230 V-3ϕ-60 Hz, or delta, voltage system is used primarily on systems that have many three-phase circuits, of 230-volt circuits, with few 120-volt circuits. The system consists of two available 120-volt legs when L_2 and L_3 are connected to a circuit with the ground. If the high leg, or L_1, is connected in a circuit with the ground, it delivers between 180 volts and 208 volts. This will damage any 115-volt load or appliance.

Detection of the 230 V-3ϕ-60 Hz system can be accomplished simply and easily by testing across any two hot legs with a voltmeter. If 230 volts are read, the system is a delta system. The high leg should be identified by an orange marking. Another method of detection is to check the voltage between each hot leg and ground with a voltmeter. If the voltage between any hot leg and ground reads between 180 volts and 208 volts, the system is a delta system.

5.6   208 VOLT-THREE-PHASE-60 HERTZ "WYE" SYSTEM

The 208 V-3ϕ-60 Hz wye voltage system is common in structures that require a large number of 120-volt circuits, such as in schools, hospitals, and office buildings. The system offers the versatility of using three-phase alternating current and the possibility of supplying many 120-volt circuits for lights, appliances, and other 120-volt equipment. Voltage ratings of some equipment are such that the equipment must be used with the right voltage system for proper operation and selection. Therefore, detection of this system is important because of the lower-voltage requirements.

This system is supplied by four wires, one ground and three hot legs, as shown in schematic form in Figure 5.14. The 208-volt system is different from the 230-volt system in that it contains no high leg. In Figure 5.14, L_1, L_2, and L_3 are the hot legs and G represents the ground. The voltage available between any two hot leg (L_1, L_2, or L_3) is 208 volts. Connections between any hot leg (L_1, L_2, or L_3) and the ground (G) provides 120 volts. As can be seen from the schematic, there are three available 120-volt power legs. This

FIGURE 5.14 Schematic for a 208 volt-three phase-60 hertz system using four wires

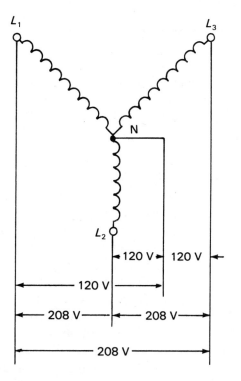

FIGURE 5.15 Schematic for the transformer hookup of a 208 V-3ϕ-60 Hz system showing the Y transformer secondary hookup

more 120-volt circuits than the 230-volt delta system.

The transformer secondary hookup of a 208 volt-three phase Y system is shown in Figure 5.15. The Y system takes its name from the letter Y, which resembles the shape of the hookup, as can be seen in Figure 5.15. All three legs of the transformer windings are equal in winding length, as shown in Figure 5.15, but in the delta system they are not. The Y transformer connection is also used on several higher-voltage systems.

Detection of the 208 V-3ϕ-60 Hz Y system is easy with the use of a voltmeter. If 208 volts are read across any two of the hot legs, or if 120 volts are read between all three hot legs and the ground, the system is a 208-volt Y system. The 208-volt Y system is a balanced system and contains no high leg that must be identified. But care should be taken to identify the ground when connections are being made.

5.7 HIGHER-VOLTAGE SYSTEMS

Higher-voltage systems are becoming increasingly popular because of their many advantages. The higher-voltage systems are used mostly on industrial structures, but in some cases they are used on commercial systems. It is necessary to understand and to be able to detect the higher-voltage system to ensure the safety of personnel and to prevent damage to the equipment.

Several high-voltage systems are available. There is a 240/480 volt-single phase system, a 240/416 volt-three phase system, and a 277/480 volt-three phase system. These systems can be identified with a voltmeter. For example, in the 277/480-volt system, the 277-volt reading is obtained between any hot leg and ground. The different systems are available with Y and delta hookups. The Y 277/480-volt system will be discussed in this section because of its popularity, but other high-voltage systems are available.

Advantages

Using higher-voltage systems has many advantages. First, all wiring is rated at 600 volts. Second, there is little difference in the switches, relays, and other electric panels used in 208-volt and 480-volt systems. Motors may be wound differently for higher voltages, but they cost a little more. The service equipment and wiring may be smaller for a 480-volt system than for a 208-volt system. This might save the consumer a great deal of money.

Disadvantages

Disadvantages to using the higher-voltage systems stem mainly from problems that would be brought about by trying to implement a common high-voltage system in this country. If the consumer needs single-phase circuits with both 120 volts and 230 volts, additional transformers must be purchased. Single-phase equipment is more common than three-phase. Equipment manufacturers would like to work toward some common voltages so they could produce fewer types of equipment. Now they must manufacture many motors with different voltage ratings to meet consumers' needs.

The 277/480-Volt System

Figure 5.16 shows a schematic layout of the 277/480-volt Y system. The system is primarily used in industry, but occasionally there are commercial applications. The system has no means of supplying 120 volts or 230 volts at single phase without the use of a separate step-down transformer.

Between any hot leg (L_1, L_2, or L_3) and ground (G), 277-volt circuits are obtained. The 277-volt circuit is commonly used in commercial and industrial lighting systems to operate fluorescent lights. Between any two hot legs

FIGURE 5.16 Schematic for a 277/480 V-3ϕ-60 Hz Y system

FIGURE 5.17 Schematic for the transformer hookup of a 277/480 V-3ϕ-60 Hz Y system

$(L_1$ and L_2, L_2 and L_3, or L_1 and $L_3)$, 480 volts can be obtained. If three-phase is desirable, it can be obtained by connections to all three hot legs. Figure 5.17 shows the completely balanced transformer layout of a 277/480 V-3ϕ-60 Hz Y system.

Detection of the 480 V-3ϕ-60 Hz system is easy with the use of a voltmeter. If 480 volts are read across any two of the hot legs, or if 277 volts are read between the hot legs and the ground, the system is a 277/480-volt Y system. The 480-volt Y system is a balanced system and contains no high leg that must be identified. But care should be taken to identify the ground when connections are being made.

SUMMARY

Alternating current is used in most heating, cooling, and refrigeration equipment because its properties give it greater flexibility than direct current. It is easier to transmit over a long distance and no expensive transmission equipment is required. It can be supplied at almost any voltage that a consumer wants.

Alternating current changes direction twice per cycle. The number of cycles per second is the frequency. Standard frequency is 60 hertz. The voltage leads the current in an AC circuit due to inductive reactance. The current leads the voltage due to capacitive reactance. The sum of the resistance and reactance in an AC circuit is called the impedance.

Voltage and current are out of phase in an AC circuit. Consequently, the power of the circuit must be calculated by using the effective wattage. The ratio between the true power and the effective power in an AC circuit is called the power factor.

An electric utility company can supply single-phase or three-phase current to structures at voltages from 208 volts to 460 volts. The generating plant supplies a high voltage that is stepped down so the consumer can use it. The common voltage characteristics are 230 V-1ϕ-60 Hz, 230 V-3ϕ-60 Hz, 208 V-3ϕ-60 Hz, and 460 V-3ϕ-60 Hz.

The 230 V-3ϕ-60 Hz, or delta, system is common throughout commercial and industrial structures. It adds efficiency to systems that have a large number of three-phase and 230-volt circuits. The system supplies three hot legs and one ground, which deliver 230 volts between any two of the three hot legs. This system has a high leg. Caution should be taken not to connect any 110-volt circuit to the high leg. Between ground and the other two hot legs, 110 volts can be obtained. Detection of the delta system is essential to prevent damage to 110-volt equipment and to ensure proper equipment selection.

The 208-volt Y system is a well-balanced three-phase system that is commonly used on structures that need many 120-volt circuits as well as three-phase or 208-volt circuits. The system does not operate motors as efficiently as the 230-volt delta system, but it is more adaptable to the balancing of each leg of a service. Much equipment is made for either 208-volt or 230-volt systems and some can be used on either system.

High-voltage systems are becoming increasingly common. In a high-voltage system smaller wire can be used, making installation of equipment less expensive. While there are several high-voltage systems in use, the 277/480-volt Y system is most common. A large number of heating, cooling, and refrigeration equipment are rated at higher voltages. Air-conditioning personnel should be able to identify the higher-voltage systems for safety and proper installation.

QUESTIONS

1. What is alternating current?

2. What is the advantage of alternating current over direct current?

3. The common frequency of alternating current used in the United States is _____.

4. What is reactance?

5. What is impedance?

6. Give a brief description of the transmission of alternating current from the generating plant to the consumer.

7. Name the four voltage-current characteristics commonly used today.

8. The phase of an alternating current is _____.

9. True or false. Inductance in an AC circuit is an effect due to the magnetic fields caused by the current flow.

10. What is the common electrical service to a residence?

11. What is the voltage range on which alternating current equipment can be operated?

12. How can a 230 V-1ϕ-60 Hz electric system be detected?

13. What is the difference between single-phase and three-phase alternating current?

14. What is the voltage of a system with a Y transformer hookup?

15. What is the voltage of a system with a delta transformer hookup?

16. What are the advantages and disadvantages of a delta electric system?

17. How can the Y and delta systems be detected?

18. What is the advantage of the 460-volt Y system over the 208-volt Y system?

19. True or false. Single-phase alternating current is common in most commercial and industrial structures.

20. Sketch the delta and Y transformer hookup arrangements.

21. What are the advantages of the 208

V-3ϕ-60 Hz system? In which types of structures is this system primarily used?

22. What is 277 volt 1ϕ-60 Hz used for?

Installation of Heating, Cooling, and Refrigeration Systems

INTRODUCTION

The proper installation of heating, cooling, and refrigeration equipment is as important as any other phase of the industry. Installation covers a broad range of subjects, but one of the most important is the electric circuit servicing the equipment and its size. Thus, industry personnel should be familiar with the structure's circuitry and circuit components.

Once the electric utility delivers the power to the structure, the consumer must bring it inside. Several kinds of electric panels are used in residences and many types are used in commercial and industrial structures to accomplish this. Hence, it is important for service and installation technicians to understand how electric power is distributed within the structure.

In this chapter we will discuss several components of the electric circuit servicing the heating, cooling, and refrigeration equipment. We will also discuss several types of electric panels that personnel may encounter on the job.

6.1 SIZING WIRE

Manufacturers usually list in the installation instructions the correct wire and fuse size. But in many cases the person responsible for the installation must calculate the wire and fuse size. The National Electrical Code governs the types and sizes of wire that can be used for a particular application and a certain amperage. The correct wire and fuse size is important to the life and efficiency of any equipment. Hence, the installation mechanic should know how to determine the size to use.

Copper is the most popular conductor in the industry. However, aluminum is used in some cases because of its low cost. Copper wire is a good conductor of electricity and has many other characteristics that make it the most popular conductor used. Copper wire bends easily, has good mechanical strength, has the ability to resist corrosion, and can be easily joined together. Aluminum conductors, on the other hand, do not have all the good characteristics of copper. Aluminum conducts electricity well enough, but problems arise because aluminum corrodes easily. Thus, aluminum wire connections have a tendency to become loose after they have corroded.

Wire Size

Standard wire size is defined in the United States by the American Wire Gauge (AWG). The American Wire Gauge lists the largest wire, 0000 (4/0), down to number 50, which is the smallest wire. In the industry, wire sizes from number 20 to number 4/0 are the most common. The most popular sizes are from number 16 to number 4.

In some cases wire larger than 4/0 is needed. The circular mil system is used for this purpose. Circular mil sizing runs from 250 MCM (MCM is the abbreviation for 1000 circular mils), which is about ½ inch in diameter, to 750 MCM, which is about 1 inch in diameter. Circular mil sizing does exist in larger sizes. Figure 6.1 is a table that gives the data on round copper wire.

Factors to Consider in Wiring

The type of insulation surrounding the conductor usually determines its application, and the amperage it can be used for. Insulation of different grades is used for different purposes. For example, insulation can be heat resistant, moisture resistant, heat and moisture resistant, or oil resistant. Figure 6.2 shows a partial table from the *National Electrical Code® (NEC)®* listing conductor application and insulations. *National Electrical Code®* and *NEC®* are registered trademarks of the National Fire Protection Association, Inc., Quincy, MA 02269.

Several factors should be considered when sizing circuit conductors. These factors are voltage drop, insulation type, enclosure, and safety. The voltage drop of a circuit, which takes into account the distance the conductors must be fed, must be calculated by the installation technician. The insulation type, the enclosure, and safety can be determined by using the tables of the *National Electrical Code®*. There are also wire-sizing tables in the *National*

American Wire Gauge (A.W.G.) Working Table (U.S. Bureau of Standards)*

| Gauge No. A.W.G. | Diameter in Mils | Cross Section | | Ohms per 1000 Ft. at 25 Deg. C. (77 Deg. F.) | Lb. per 1000 Ft. |
		Circular Mils	Square Inches		
0000	460	212000	.166	.0500	641
000	410	168000	.132	.0630	508
00	365	133000	.105	.0795	403
0	325	106000	.0829	.100	319
1	289	83700	.0657	.126	253
2	258	66400	.0521	.159	201
3	229	52600	.0413	.201	159
4	204	41700	.0328	.253	126
5	182	33100	.0260	.320	100
6	162	26300	.0206	.403	79.5
7	144	20800	.0164	.508	63.0
8	128	16500	.0130	.641	50.0
9	114	13100	.0103	.808	39.6
10	102	10400	.00815	1.02	31.4
11	91	8230	.00647	1.28	24.9
12	81	6530	.00513	1.62	19.8
13	72	5180	.00407	2.04	15.7
14	64	4110	.00323	2.58	12.4
15	57	3260	.00256	3.25	9.86
16	51	2580	.00203	4.09	7.82
17	45	2050	.00161	5.16	6.20
18	40	1620	.00128	6.51	4.92
19	36	1290	.00101	8.21	3.90
20	32	1020	.000802	10.4	3.09
21	28.5	810	.000636	13.1	2.45
22	25.3	642	.000505	16.5	1.94
23	22.6	509	.000400	20.8	1.54
24	20.1	404	.000317	26.2	1.22
25	17.9	320	.000252	33.0	0.970
26	15.9	254	.000200	41.6	0.769
27	14.2	202	.000158	52.5	0.610
28	12.6	160	.000126	66.2	0.484
29	11.3	127	.0000995	83.5	0.384
30	10.0	101.0	.0000789	105	0.304
31	8.9	79.7	.0000626	133	0.241
32	8.0	63.2	.0000496	167	0.191
33	7.1	50.1	.0000394	211	0.152
34	6.3	39.8	.0000312	266	0.120
35	5.6	31.5	.0000248	336	0.0954
36	5.0	25.0	.0000196	423	0.0757
37	4.5	19.8	.0000156	533	0.0600
38	4.0	15.7	.0000123	673	0.0476
39	3.5	12.5	.0000098	848	0.0377
40	3.1	9.9	.0000078	1070	0.0299

FIGURE 6.1 Data on round copper wire *(Courtesy of Continental Wire & Cable, Division of the Anaconda Company)*

Trade Name	Type Letter	Max. Operating Temp.	Application Provisions	Insulation	AWG or kcmil	Thickness of Insulation	Mils	Outer Covering****
Heat-Resistant Thermoplastic	THHN††	90°C 194°F	Dry and damp locations.	Flame-Retardant, Heat-Resistant Thermoplastic	14-12 .... 10 .... 8-6 .... 4-2 .... 1-4/0 .... 250-500 .... 501-1000		15 20 30 40 50 60 70	Nylon jacket or equivalent
Moisture- and Heat-Resistant Thermoplastic	THHW	75°C 167°F 90°C 194°F	Wet location Dry location	Flame-Retardant, Moisture- and Heat-Resistant Thermoplastic	14-10 .... 8-2 .... 1-4/0 .... 213-500 .... 501-1000		45 60 80 95 110	None
Moisture- and Heat-Resistant Thermoplastic	THW††, †††	75°C 167°F 90°C 194°F	Dry and wet locations. Special applications within electric discharge lighting equipment. Limited to 1000 open-circuit volts or less. (Size 14-8 only as permitted in Section 410-31.)	Flame-Retardant, Moisture and Heat-Resistant Thermoplastic	14-10 .... 8-2 .... 1-4/0 .... 213-500 .... 501-1000 .... 1001-2000		45 60 80 95 110 125	None
Moisture- and Heat-Resistant Thermoplastic	THWN ††, †††	75°C 167°F	Dry and wet locations.	Flame-Retardant, Moisture- and Heat-Resistant Thermoplastic	14-12 .... 10 .... 8-6 .... 4-2 .... 1-4/0 .... 250-500 .... 501-1000		15 20 30 40 50 60 70	Nylon jacket or equivalent
Moisture-Resistant Thermoplastic	TW††	60°C 140°F	Dry and wet locations.	Flame-Retardant, Moisture-Resistant Thermoplastic	14-10 .... 8 .... 6-2 .... 1-4/0 .... 213-500 .... 501-1000 .... 1001-2000		30 45 60 80 95 110 125	None
Underground Feeder & Branch-Circuit Cable-Single Conductor. (For Type UF cable employing more than one conductor, see Article 339.)	UF	60°C 140°F **75°C 167°F	See Article 339.	Moisture-Resistant Moisture- and Heat-Resistant	14-10 .... 8-2 .... 1-4/0		*60 *80 *95	Integral with insulation

*Includes integral jacket.

**For ampacity limitation, see Section 339-5.

****Some insulations do not require an outer covering.

††Insulation and outer coverings that meet the requirements of flame-retardant, limited smoke and are so listed shall be permitted to be designated limited smoke with the suffix /LS after the Code type designation.

†††Listed wire types designated with the suffix -2 such as RHW-2 shall be permitted to be used at a continuous 90°C operating temperature wet or dry. Ampacities of these wire types are given in the 90°C column in the appropriate ampacity table.

FIGURE 6.2 Table of conductor application and insulation *(Reprinted with permission from NFPA 70-1990, the National Electrical Code®. Copyright © 1990, National Fire Protection Association, Inc., Quincy, MA. This reprinted material is not the complete and official position of the NFPA on the referenced subject, which is represented only by the standard in its entirety.)*

Electrical Code® that give the allowable amperage for both aluminum and copper conductors. The *National Electrical Code®* is considered a guide to safe wiring procedures. The code does not ensure that all systems following its procedures will be good systems, however. That is the responsibility of the designer.

Voltage Drop

Voltage drop in a conductor is of prime importance when sizing wire. Any voltage that drops between the supply and the equipment is lost to the equipment. If the voltage drop is large enough, it will seriously affect the operation of the equipment. But even a small voltage drop is detrimental to the equipment. The voltage drop can be easily measured when the equipment is operating by reading the voltage at the supply and subtracting from that the voltage read at the equipment. If we read 240 volts at the supply and 210 volts at the equipment when it is operating, then there is a voltage drop of 30 volts in the circuit. The allowable drop of most equipment is 10%.

Wire-Sizing Charts

Figure 6.3 shows the table from the *National Electrical Code®* for the allowable ampacities (amperages) of a conductor for copper and aluminum wire. The wire-sizing charts from the *NEC®* are usually accurate for sizing electric conductors unless the circuit is extremely long. For example, if a 5-ton condensing unit pulls 24 amperes, the service mechanic should add 25% of the full-load amperage to the total, which would be 6 amperes. This gives a total of 30 amperes for the wire-sizing data. From the table in Figure 6.3, we see that the circuit would require a copper conductor of No. 10 TW copper wire. Also from the table, we see the circuit would require a No. 8 TW aluminum conductor.

Figure 6.4 shows the manufacturer's electrical data for a piece of equipment. The wire sizes are given in the first and third columns. Figure 6.5 shows the table of electrical characteristics of a specific model of equipment, but in this table the exact wire sizes are not given. However, the wire size amperages are given (the column headed MWA). The installation mechanic would use the wire size amperage column and the *NEC®* wire charts to size the wire.

Many items in the *NEC®* cover the sizing of electric conductors. Most technicians use the *NEC®* for reference. The preceding examples were given as a guide to show you how to use the NEC charts.

ARTICLE 310—CONDUCTORS FOR GENERAL WIRING   70-161

**Table 310-16. Ampacities of Insulated Conductors
Rated 0-2000 Volts, 60° to 90°C (140° to 194°F)
Not More Than Three Conductors in Raceway or Cable or Earth
(Directly Buried), Based on Ambient Temperature of 30°C (86°F)**

Size	Temperature Rating of Conductor, See Table 310-13.								Size
	60°C (140°F)	75°C (167°F)	85°C (185°F)	90°C (194°F)	60°C (140°F)	75°C (167°F)	85°C (185°F)	90°C (194°F)	
AWG kcmil	TYPES †TW, †UF,	TYPES †FEPW, †RH, †RHW, †THHW, †THW, †THWN, †XHHW, †USE, †ZW	TYPE V	TYPES TA, TBS, SA SIS, †FEP, †FEPB, †RHH, †THHH, †THHW, †XHHW	TYPES †TW, †UF	TYPES +RH, †RHW, †THHW, †THW, †THWN, †XHHW, †USE	TYPE V	TYPES TA, TBS, SA, SIS, †RHH, †THHW, †THHN, †XHHW	AWG kcmil
	COPPER				ALUMINUM OR COPPER-CLAD ALUMINUM				
18	....	....	....	14	....	....	....	....	....
16	....	....	18	18	....	....	....	....	....
14	20+	20+	25	25+	....	....	....	....	....
12	25+	25+	30	30+	20+	20+	25	25+	12
10	30	35+	40	40+	25	30+	30	35+	10
8	40	50	55	55	30	40	40	45	8
6	55	65	70	75	40	50	55	60	6
4	70	85	95	95	55	65	75	75	4
3	85	100	110	110	65	75	85	85	3
2	95	115	125	130	75	90	100	100	2
1	110	130	145	150	85	100	110	115	1
1/0	125	150	165	170	100	120	130	135	1/0
2/0	145	175	190	195	115	135	145	150	2/0
3/0	165	200	215	225	130	155	170	175	3/0
4/0	195	230	250	260	150	180	195	205	4/0
250	215	225	275	290	170	205	220	230	250
300	240	285	310	320	190	230	250	255	300
350	260	310	340	350	210	250	270	280	350
400	280	335	365	380	225	270	295	305	400
500	320	380	415	430	260	310	335	350	500
600	355	420	460	475	285	340	370	385	600
700	385	460	500	520	310	375	405	420	700
750	400	475	515	535	320	385	420	435	750
800	410	490	535	555	330	395	430	450	800
900	435	520	565	585	355	425	465	480	900
1000	455	545	590	615	375	445	485	500	1000
1250	495	590	640	665	405	485525	545	1250	
1500	520	625	680	705	435	520	565	585	1500
1750	545	650	705	735	455	545	595	615	1750
2000	560	665	725	750	470	560	610	630	2000

AMPACITY CORRECTION FACTORS									
Ambient Temp. °C	For ambient temperatures other than 30°C (86°F), multiply the ampacities shown above by the appropriate factor shown below.								Ambient Temp. °F
21-25	1.08	1.05	1.04	1.04	1.08	1.05	1.04	1.04	70-77
26-30	1.00	1.00	1.00	1.00	1.00	1.00	1.00	1.00	79-86
31-35	.91	.94	.95	.96	.91	.94	.95	.96	88-95
36-40	.82	.88	.90	.91	.82	.88	.90	.91	97-100
41-45	.71	.82	.85	.87	.71	.82	.85	.87	106-113
46-50	.58	.75	.80	.82	.58	.75	.80	.82	115-122
51-55	.41	.67	.74	.76	.41	.67	.74	.76	124-131
56-60	....	.58	.67	.71	....	.58	.67	.71	133-140
61-70	....	.33	.52	.58	....	.33	.52	.58	142-158
71-80	....	....	.30	.41	....	....	.30	.41	160-176

†Unless otherwise specifically permitted elsewhere in this Code, the overcurrent protection for conductor types marked with an obelisk (†) shall not exceed 15 amperes for 14 AWG, 20 amperes for 12 AWG, and 30 amperes for 10 AWG copper; or 15 amperes for 12 AWG and 25 amperes for 10 AWG aluminum and copper-clad aluminum after any correction factors for ambient temperature and number of conductors have been applied.

FIGURE 6.3 Table for wire sizing *(Reprinted with permission from NFPA 70-1990, the National Electrical Code®. Copyright © 1990, National Fire Protection Association, Inc., Quincy, MA. This reprinted material is not the complete and official position of the NFPA on the referenced subject, which is represented only by the standard in its entirety.*

BRANCH CIRCUIT

Power Wire Size (AWG)	Max Ft Wire	Gnd Wire Size† (AWG)	Max Fuse Amps
10	40	10	35
10	35	10	40
8	43	10	50
8	35	10	60
8	44	10	60
6	56	8	70
12	32	12	30
10	47	10	35
10	36	10	45
8	45	10	50
8	56	10	50
8	47	10	60
10/12	50/42	10/12	35/30
10/10	45/60	10/10	35/30
10	53	10	40
10	45	10	45
10	68	10	35
10	57	10	40
14	79	12	20
14	67	12	20

†Required when using nonmetallic conduit.

FIGURE 6.4 Table of recommended wire sized for an air conditioner *(Reproduced by permission of Carrier Corporation. Copyright © 1977 Carrier Corporation, Syracuse, NY)*

Calculating Voltage Drop

The table previously shown in Figure 6.3 does not consider voltage drop, which must be calculated for long circuits. To calculate the voltage drop in a conductor, you must know the number of feet of wire that is used. The following formula is used to calculate voltage drop:

$$E = IR$$

For example, if No. 12 TW wire is to run 500 feet, what is the voltage drop over this distance when the supply voltage is 240 volts? First, we find the resistance of No. 12 wire from Figure 6.1. It is 1.6 ohms per 1000 feet. From Figure 6.3 we find the ampacity of No. 12 TW wire. It is 20 amperes. Substituting in the formula $E = IR$, we find $E = 20 \times .8$ (1.6 ohms per 1000 feet = .8 ohm per 500 feet). The voltage drop over 500 feet is 16 volts.

UNIT MODEL 38AD	V/PH Nom	V/PH Min-Max	MWA	ICF	FU	COMPR FLA	COMPR LRA	FANS FLA (ea)*
410	208/3	187-229	68	198	80	49.3	191	3.2
012 510	230/3	198-254	62	179	70	44.3	172	3.2
610	460/3	414-506	31	89	35	22.2	86	1.6
110	575/3	518-632	24	71	30	17.9	69	1.0
400	208/3	187-229	86	184	100	63.6	173/266	3.2
014 500	230/3	198-254	78	166	90	57.2	160/240	3.2
600	460/3	414-506	39	123	45	28.6	120	1.6
100	575/3	518-632	32	99	35	22.9	96	1.0
400	208/3	187-229	86	184	100	64.0	173/266	3.2
016 500	230/3	198-254	78	166	90	58.0	160/249	3.2
600	460/3	414-506	39	123	45	29.0	96	1.6
100	575/3	518-632	32	99	40	23.0	96	1.0

☐ Values are for part-winding start.

FLA – Full Load Amps
FU – Fuse (max allowable amps; dual element)
ICF – Max Instantaneous Current Flow (during start-up; sum of compressor LRA plus FLA of all other motors in the unit).
LRA – Locked Rotor Amps
MWA – Minimum Wire Amps per NEC

*Units 38AD012 and 014 have two; unit 38AD016 has three.

FIGURE 6.5 Table of electrical characteristics of a large condensing unit *(Reproduced by permission of Carrier Corporation. Copyright © 1977 Carrier Corporation, Syracuse, NY)*

Subtracted from the supply voltage of 240 volts, this gives 224 volts supplied to the equipment.

The formula

$$E_d = \frac{22 \times D \times I}{\text{cmil}}$$

can also be used, where E_d is the voltage drop, D is the distance, I is the current, and cmil is circular mils.

In most cases, the wiring that is run to heating, cooling, or refrigeration equipment will not exceed 75 feet to 100 feet. Then the wire size can be read directly from the table in Figure 6.3 because the voltage drop can be ignored on short-distance circuits. (There is a correction factor in the NEC tables if the temperature exceeds 30° C.)

6.2 DISCONNECT SWITCHES

All heating, cooling, and refrigeration equipment should have some means for disconnecting the power supply at the equipment. Some equipment has a built-in method for disconnecting the power, such as a circuit breaker or fuse blocks. However, in most cases a disconnecting device must be supplied and installed by an electrician or the installation mechanic. Disconnect switches are relatively simple and easy to install once the correct selection is made.

A disconnect switch is a two- or three-pole switch mounted in an enclosure. The switch can be purchased with or without a space for fuses. Disconnect switches in most cases have a grounding lug mounted in the enclosure. The switches can have different arrangements. For example, a four-pole-three-fuse switch would be used on a three-phase circuit and would have a grounding lug. A three-pole-three-fuse switch would be used for three-phase circuits and would not have a ground lug. A three-pole-two-fuse switch would be used on single-phase circuits and would have a ground lug. Figure 6.6 shows the schematics of two disconnect switches and Figure 6.7 shows these switches.

Disconnect switches can be purchased for general duty or heavy duty. The heavy-duty disconnect switch would be installed for equipment requiring frequent use. The general-duty switch would be used for equipment requiring infrequent use.

Enclosures

The type of enclosure that the disconnect switch mounts in is determined by the conditions existing in the area of installation. A general type of enclosure could only be used where there were no problems of moisture, dust, or explosive fumes. A watertight disconnect enclosure could be used in areas of moisture but not where dust or explosive fumes exist. An explosion proof

   (a) 3-wire—2-fuse disconnect   **(b) 4-wire—3-fuse disconnect**

FIGURE 6.6 Schematic diagram of two fusible disconnect switches

(a) **4-wire—3-fuse disconnect**   (b) **3-wire—2-fuse disconnect**

FIGURE 6.7 Disconnect switches

enclosure could be used in any location, but it is much more expensive than other enclosures and hence is not used without reason.

Fusible and Nonfusible Switches

The purpose of a disconnect switch can be twofold. First, it can be used as a means of disconnecting the supply power going to the equipment. Second, it can be used as a safety device when fused correctly. If the only purpose of a disconnect switch is to break the power supply, then a nonfusible disconnect switch should be used. But if a means of protection for the wire or equipment is needed, a fusible disconnect switch should be used with the proper size fuses. Most equipment manufacturers will give the fuse sizes needed in the installation instructions. If fuse sizes are not given, the *National Electrical Code* should be consulted.

The selection of a fusible disconnect switch is determined by duty, enclosure type, and size. Fuses are designed so that one size covers several different ampacities. The same size fuse can be purchased to cover from 1 to 30 amperes, from 30 to 60 amperes, from 70 to 100 amperes, and from 100 to

200 amperes. There are larger sizes available, but they are not used frequently. Disconnect switches are rated 30 amperes, 60 amperes, 100 amperes, 200 amperes, 400 amperes, and 600 amperes. A 30-ampere disconnect switch would be used for any load from 1 ampere to 30 amperes. A 200-ampere disconnect switch can be used with fuses from 100 amperes to 200 amperes. Other determining factors of the switches can easily be selected from a manufacturer's catalog.

Figure 6.8 shows a disconnect switch installed on a piece of equipment. In the installation of a disconnect switch, it is imperative that the wiring connections be made tightly. The line voltage wires should always be connected at the top of the enclosure. The load or wires supplying the unit should be connected to the load side or bottom of the fuses. These connections can be seen in Figure 6.8. At no time should the wires be installed in an enclosure without the proper connection.

6.3 FUSIBLE LOAD CENTERS

Fusible load centers, or breaker panels, are electric panels that supply the circuits in a structure with power and protect those circuits with fuses. Fig-

FIGURE 6.8 Disconnect switch installed on equipment

FIGURE 6.9 Fusible load center

ure 6.9 shows a typical fusible load center. Fusible load centers were popular until about 1968, when breaker panels gained prominence in the market. Air-conditioning mechanics often find themselves working on fusible load centers when the owner of an older home decides to have air conditioning installed. Many fusible load centers are still in existence, so mechanics must understand them and know how to make the correct connections. In many cases, it is impossible to add a 230-volt circuit to a fusible load center, so the service technician must know when to recommend installing a new breaker panel.

One of the earliest fusible load centers is shown in schematic form in Figure 6.10. The panel had a 60-ampere lighting main, which means that the total lighting or 110-volt circuits in the residence could not exceed 60 amperes on the eight circuits. There was also a 60-ampere-230-volt circuit for a kitchen stove. The addition of any electric circuit to the 60-ampere fusible load center is almost impossible because of its construction and its low capacity. Most modern residences require a 150-ampere or 200-ampere service. Fusible load centers built after about 1958 are capable of handling the load.

Many fusible load centers have been installed in residences that are much better and more versatile than the original 60-ampere centers. Many are designed with a lighting main and a capacity to take a number of additional 230-volt circuits.

The newest fusible load centers are built with the capacity of taking additional fuse blocks in the main lugs, which supply power to the entire panel, at any time if a space is available. This capacity allows the electrician to

FIGURE 6.10 Schematic of an early fusible load center

install the additional circuits required for any particular structure. Figure 6.11 shows this panel and its schematic. It is relatively easy to add a circuit to the late-model panels by merely inserting the fuse blocks and screwing them in.

6.4 BREAKER PANELS

Breaker panels are usually installed in residences and industrial buildings today. Breakers are devices that detect any overload above their rating in a circuit and open the circuit automatically. The breaker must then be reset manually.

(a) Schematic

(b) Load center

FIGURE 6.11 Modern fusible load center

Breakers can be obtained for almost any application, no matter how large or small. Breakers are made with one, two, or three poles. The poles denote how many hot legs are being fed from the breaker to the appliance. A one-pole breaker supplies one hot leg and makes up a 110-volt–single-phase circuit. A two-pole breaker supplies two hot legs and makes a 230-volt–single-phase circuit. A three-pole breaker supplies three hot legs and makes up a three-phase circuit at the supplied voltage. Breakers are supplied in amperage ratings of one ampere to several hundred amperes, depending on the application. Figure 6.12 shows one-pole, two-pole, and three-pole breakers.

Construction

Breaker panels are built in several different forms, but they all serve the same purpose. Different manufacturers build breaker panels that are similar in design, but the breakers from different manufacturers do not usually fit each others' panels. However, several manufacturers' equipment does interchange.

Breaker panels are built with or without main breakers. A main breaker installed in the breaker panel provides a main switch in the panel and adds a means of overload protection for the entire panel. A breaker panel can be obtained with main lugs and no main breaker, but the breaker should have some means of protection. Breaker panels are rated by how many amperes the main lugs can carry and by the rating of the main breaker. Figure 6.13

FIGURE 6.12 (a) One-pole breaker (b) Two-pole breaker (c) Three-pole breaker

FIGURE 6.13 Breaker panel showing main lugs and main breaker shows the main lugs of a breaker panel with a main breaker.

Breaker panels are built for use with single-phase or three-phase systems and for 250 volts or 600 volts. The breaker panel in the average residence is rated at 150 amperes or 200 amperes and is a general-duty type. The breakers snap into a residential panel (as they do in some commercial and industrial panels). Figure 6.14 shows a residential breaker panel and its schematic.

The commercial and industrial breaker panel can meet almost any specifications that the consumer requires. The commercial and industrial panels are built for more rugged duty than are the usual residential panels. Most commercial and industrial panels use breakers that attach to main lugs with screws. Figure 6.15 shows a typical breaker panel, along with its schematic, used in commercial and industrial applications.

Installation

Installation of a breaker into a breaker panel causes little or no trouble. Breakers connect in some panels by attaching to the main lugs with clips. Other breakers are attached to the main lugs by screws. You should be familiar with breakers and breaker panels to order the correct breaker for a particular use.

Single Main
1 Phase, 3 Wire

S/N

FEQ12MB-100A
WFEQ12MB-100A
12 Poles max

(a) Schematic　　　**(b) Panel**

FIGURE 6.14 Residential breaker panel

S/N

EQ 430MBD—400A
EQ 430MBE—600A
30 poles max.

(a) Schematic　　　**(b) Panel**

FIGURE 6.15 Industrial breaker panel *(Courtesy of Gould I-T-E Electrical Products)*

In most cases, breaker panels have several open spaces. If a situation arises where there is no spare opening, a smaller breaker can replace the larger standard breakers. Most manufacturers provide the smaller breaker. Figure 6.16 shows the comparison between the standard breaker and the smaller compact breaker.

Figure 6.17 shows a breaker being installed in a breaker panel. Care should be taken to ensure the breaker is correctly aligned with the proper clip in the panel. When installing a breaker that is attached to the lug with a screw, make certain that the connection is tight. Before the work of installing a breaker in a panel is begun, it is recommended that the power be cut off.

On rare occasions circuit breakers are faulty. Either they cannot be reset or they open the circuit on a lower amperage than the rating. If either condition exists, the breaker must be replaced. A breaker can be checked for resetting if a voltage reading is taken between the ground and the breaker. If voltage is available to the load side of the breaker under load, the breaker is good. In rare cases, a breaker could be stuck in the closed position, and if so, it should be replaced. The breaker that trips at a lower-than-rated amperage should be checked with an actual ampere reading of the circuit.

FIGURE 6.16 Standard breaker and small replacement breaker

FIGURE 6.17 Breaker being installed in a breaker panel

6.5 DISTRIBUTION CENTERS

Distribution centers are designed to distribute the electrical supply to several places in a large structure. Their use is largely confined to commercial and industrial applications. Figure 6.18 shows schematically the layout of a distribution system in a structure. In Figure 6.18, panel A is the main distribution point between the electrical service and other panels and the heating and air-conditioning equipment.

Figure 6.19 shows a modern distribution panel. In a very large structure, it would not be unusual to find several distribution centers. A distribution center often saves a great deal of money because many circuits are shortened and only the main circuit is lengthy. Distribution centers can be of the fusible or circuit-breaker design. The fusible design is the more popular mainly because of its moderate cost. Larger breakers are extremely expensive and hence are not used often.

FIGURE 6.18 Schematic of the electrical distribution system of a large structure

FIGURE 6.19 Modern distribution panel

SUMMARY

For proper installation and maintenance of heating, cooling, and refrigeration equipment, industry personnel should be familiar with the electric circuits and circuit components of a structure. The life and safety of the equipment depend on the use of the correct size of wiring. Voltage drop, insulation type, enclosure, and safety are the determining factors in wire sizing. Manufacturers usually provide instructions on the size of wire to use with their equipment, but in some cases the mechanic must calculate the size. The *National Electrical Code®* is a guide that should be used in sizing wire properly.

Various types of electric panels are used in the industry. The disconnect switch, sometimes called the safety switch, is commonly used on equipment, along with another electric panel. In most structures, the electric panels are not located close enough to the equipment to be considered safe for disconnecting the equipment. Therefore, in most cases disconnect switches should be installed on or close to the equipment.

The fusible load center or breaker panel is used in most residences. The breaker panel is also popular in commercial and industrial panels. Breakers are designed to trip or break the circuit on an overload. Some breakers clip to the main lugs; others screw to them.

The distribution center is used in commercial and industrial plants as a means of distributing power to other electric panels in the structure. Service personnel should be familiar with power distribution because of the many voltage ranges of modern equipment. The mechanic or installer must be able to pick up power out of any type of electric panel. In many areas, industry personnel are responsible for the total installation of equipment, including power wiring. In other areas, power wiring must be done by an electrician.

QUESTIONS

1. What size wire could be used for a circuit requiring 100 amperes?

2. What are the four factors that should be considered in sizing circuit conductors?

3. What is a disconnect switch?

4. What are the three basic types of disconnect switch enclosures?

5. What is a fusible load center?

6. What is a circuit breaker?

7. How can a faulty circuit breaker be detected?

8. How are circuit breakers attached to the lugs of an electric panel?

9. What is a distribution center and what is its primary use?

10. The most popular conductor used in the industry is _____.

11. True or false. The American Wire Gauge lists the standard wire sizes used in the United States today.

12. What is the major disadvantage of aluminum wiring?

13. How would you determine the voltage drop in a circuit?

14. For wire-sizing problems the service technician should consult the _____.

15. What is the purpose of a disconnect switch?

16. True or false. Fusible load centers are usually installed in new homes today.

17. Breakers are made with _____ poles. The poles denote _____.

18. What is the purpose of a main lug in a breaker panel?

19. What would you do in a situation where there is no spare opening in a breaker panel for installation of equipment?

20. If a 10-TW wire is 750 feet, what is the voltage drop in 230 volts?

21. What would the supply voltage be for a 100-foot 4/0 wire if 150 amperes are used on 230 volts?

LAB SECTION REFERENCE

For experiments and activities dealing with material covered in this chapter, refer to Lab Number 5, Sizing Wire for Air-Conditioning Condensing Units.

7

Basic Electric Motors

INTRODUCTION

The electric motor changes electric energy into mechanical energy. Motors are used to drive compressors, fans, pumps, dampers, and any other device that needs energy to power its movement.

There are many different types of electric motors with different running and starting characteristics. Most single-phase motors are designed and used according to their running and starting torque. **Torque** is the strength that a motor produces by turning, either while starting or running. This chapter covers most types of motors available today and how they are used in the heating, cooling, and refrigeration industry.

We begin our study with a discussion of magnetism, an effect that is needed to operate motors, relays, contactors, and other electric devices.

7.1 MAGNETISM

Magnetism is the physical phenomenon that includes the attraction of an object for iron and is exhibited by a permanent magnet or an electric current. Magnetism is produced in many different ways, but regardless of how it is produced, the effect is basically the same. The magnetic field of the earth, for example, is the same as the magnetism in a horseshoe magnet, the magnetism produced by a transformer, and the magnetism produced by an electromagnet. A good example of magnetism is the ability of a horseshoe magnet to pick up articles made of iron. The most common example of magnetism is the reaction of a compass to the earth's magnetic field.

All magnets have two poles, a north pole and a south pole. If the north pole of a bar magnet is brought close to the north pole of another bar magnet, they will repel each other, as shown in Figure 7.1. If the south pole of one bar magnet is brought close to the north pole of another bar magnet,

FIGURE 7.1 Repulsion of like poles of two bar magnets

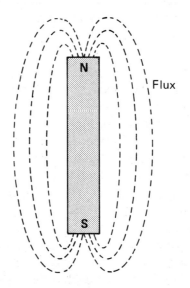

FIGURE 7.2 Attraction on unlike poles of two bar magnets

they will attract each other and come together, as shown in Figure 7.2. Therefore, like poles of magnets repel each other and unlike poles attract.

Magnetic Field

The magnetic lines of force of a magnet that flow between the north and south poles are called **flux**. These lines of force are shown in Figure 7.3. The

FIGURE 7.3 Magnetic field of a bar magnet

area that the magnetic force operates in is called a **magnetic field**. Magnetic fields can flow through materials, depending on the strength of the magnetic field. A magnetic field is best conducted through soft iron. That is why certain parts of motors and other electric devices are made of soft iron.

Induced Magnetism

Induced magnetism is created when a piece of iron is placed in a magnetic field. The important fact to remember about a magnetic field is that the closer an object is to the magnet, the stronger the magnetic field is on that object. Therefore, if we insert an iron bar within two or three inches of a magnetic field, we induce a stronger field than if we placed the bar six inches from the field.

Two types of magnets are in use today: the permanent magnet and the electromagnet. The **permanent magnet** is a piece of magnetic material that has been magnetized and can hold its magnetic strength for a reasonable length of time. The permanent magnet must be made of a magnetic material, such as iron, nickel, cobalt, or chromium. Some nonmagnetic materials such as glass, rock, wood, paper, and air cannot be made magnetic but can be penetrated by a magnetic field.

The **electromagnet** is a magnet produced by electricity. When an electron flow is in a conductor, a magnetic field is created around the conductor, as shown in Figure 7.4. The larger the electron flow, the stronger the magnetic field. Therefore, if we take an iron core and wind a current-carrying conductor around it, the iron core will become a magnet, as shown in Figure 7.5. The electron flow and the number of turns of the conductor around the

FIGURE 7.4 Magnetic field created around a current-carrying conductor

FIGURE 7.5 Magnetic field of an iron core when a current-carrying conductor is wound around the core

FIGURE 7.6 An electromagnet used as a solenoid in a contactor

core determine the strength of an electromagnet. Figure 7.6 shows an electromagnet that is used as a solenoid in a contactor.

Magnetism is important in the heating, cooling, and refrigeration industry because of its many uses in the operation of electric devices. Motors require magnetism to create a rotating motion. Relays and contactors use magnetism to open and close a set of contacts. All the devices discussed in this chapter use magnetism in some way.

7.2 BASIC ELECTRIC MOTOR

Electric motors are common electric devices in the heating, cooling, and refrigeration industry. Motors are used to create a rotating motion and drive components that need to be turned. Motors power compressors, pumps, fans, timers, and any other device that must be driven with a rotating motion.

In an electric motor, electric energy is changed to mechanical energy by magnetism, which causes the motor to turn. The method by which magnetism causes motors to rotate uses the principle that like poles of magnets repel and unlike poles attract. Suppose a simple magnet is placed on a pivot and used as a **rotor** (the rotating part of an electric motor) and a horseshoe magnet is used as a **stator** (the stationary part of a motor), as shown in Figure 7.7(a). Movement will be obtained by the repulsion and attraction of the poles of a magnet. The rotor would turn until the unlike poles are attracted to each other, as shown in Figure 7.7(b).

To make an electric motor move continuously, we must have a rotating magnetic field, which is produced by the reversal of the poles, or the polar-

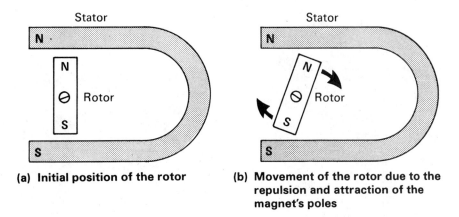

(a) **Initial position of the rotor**   (b) **Movement of the rotor due to the repulsion and attraction of the magnet's poles**

FIGURE 7.7 A simple electric motor

ity, in the rotor or stator. An alternating current of 60 hertz changes direction 120 times per second. Therefore, the current would change the polarity of the stator poles on each reversal of current. If the rotor has a permanent polarity, as shown in Figure 7.8, then the changes of polarity in the stator would cause the rotor to move. Therefore, if alternating current changes direction, causing a polarity change, 120 times a second, then the motor will turn in a continuous motion because the poles of the stator will be continuously repelling and attracting the permanent poles of the rotor. Figure 7.8 shows the motor in one complete cycle of current, or 1/60 second. The movement of the motor is caused by the magnetic field of the stator as it rotates through its alternations of current. Figure 7.9 shows the coils of an ordinary electric motor and its windings. The coils of the motor are a part of the stator.

Two-pole motor, 3600 rpm

FIGURE 7.8 Complete cycle of operation of an electric motor

FIGURE 7.9 Windings of an electric motor

FIGURE 7.10 Squirrel cage rotor

In motors, the rotor is not a permanent magnet, as we stated in the previous explanation. The **squirrel cage rotor** shown in Figure 7.10 is the most commonly used rotor today. The squirrel cage rotor derives its name from its cagelike appearance. In the squirrel cage rotor, copper or aluminum bars are evenly spaced in the steel portion of the rotor and connected by an aluminum or copper end ring. The squirrel cage rotor produces an inductive magnetic field within itself when the stator is energized.

The most common motors operate much like a transformer, with the stator being the primary magnetic field and the rotor being a movable secondary magnetic field. The rotor will have magnetism induced into it from the stator and its magnetic poles will be permanent. The magnetic poles of the stator are moving at the rate of the alternations of the current.

7.3 TYPES OF ELECTRIC MOTORS

The industry uses all kinds of AC motors to rotate the many different devices that require rotation in a complete system. Different motors are needed for different tasks because all motors do not have the same running and starting characteristics. This fact, along with the increased cost of stronger motors, allows the industry to use the right motor for the right job. For example, compressors require a motor with a high starting torque and a good running efficiency. Small propeller fans use motors with a low starting torque and average running efficiency.

Motor Strength

The starting methods, or strength, are generally used to classify motors into types. Motors are selected mainly because of the starting torque (power) required for the motor to perform its function. Five general types of motors are used in environmental systems: shaded-pole, split-phase, permanent split-capacitor, capacitor-start-capacitor-run, capacitor-start, and three-phase. There are others, such as the repulsion-start-induction-run and series motors. However, these are outdated or not commonly used in the industry. The starting torques of the five general types of induction motors, expressed as a percentage of their running torque, are as follows: shaded-pole, 100%; split-phase, 200%; permanent split-capacitor, 200%; capacitor-start-capacitor-run, 300%; and three-phase, 600%.

Open and Enclosed Motors

Two other motor types are the open motor and the enclosed motor. The open motor, shown in Figure 7.11, has a housing and is used to rotate a device such as a fan or a pump that is itself not enclosed in any type of housing. The enclosed motor is housed within some type of shell. The most common enclosure of a motor is a completely sealed hermetic compressor, as shown in Figure 7.12. Any starting apparatus used on an enclosed motor must be mounted outside the enclosure. The starting apparatus of an open motor is usually mounted within the motor itself.

In the following sections, we will discuss the five basic motor types in detail.

FIGURE 7.11 Open motor

FIGURE 7.12 Enclosed motor used in hermetic compressors

7.4 SHADED-POLE MOTORS

Most single-phase induction motors require a starting winding to create a starting torque that enables the motor to start. In most cases, the starting winding is located 90 electrical degrees from the main winding. A **shaded-pole motor** uses a shaded pole made of a closed turn of a heavy copper wire banded around a section of each stator pole. A shaded-pole motor is shown in Figure 7.13. Shaded-pole motors are used when very small starting and running torques are required, such as in a furnace fan, a small condensing unit fan, and an open-type propellor fan. These motors are easily stalled, but in most cases, because of the small locked rotor amperes (i.e., the current draw of the motor when power is applied but the motor does not turn), they can stall and still not burn out the windings.

Operation

Figure 7.14 shows the stator of a shaded-pole motor. At one side of each pole, a small groove has been cut into the stator and banded by a solid copper wire or band, as shown in Figure 7.15.

When the shaded-pole motor is starting, a current is induced into the shaded pole from the main windings. The shaded poles produce a magnetic field that is out of phase with the magnetic field of the main winding, and a rotating magnetic field is produced that is sufficient to give the desired starting torque. When the motor approaches full speed, the effect of the shaded pole is negligible. The rotation of the shaded-pole motor is from the unshaded pole to the shaded pole.

Figure 7.16 shows the schematic diagram of a single-speed, shaded-pole motor. A single-speed, shaded-pole motor has only one winding, with the

FIGURE 7.13 Shaded-pole motor

FIGURE 7.14 Stator of a shaded-pole motor

FIGURE 7.15 Shaded pole

FIGURE 7.16 Schematic wiring diagram of the shaded-pole motor

exception of the shaded pole, and is relatively simple. Figure 7.17 shows the schematic diagram of a three-speed, shaded-pole motor. The main winding is the speed winding for high-speed operation. For medium-speed operation the main winding is put in series with the medium-speed winding, which increases the number of poles and produces fewer revolutions per minute. For low-speed operation the main winding is put in series with the medium- and low-speed windings, which increases the number of poles in the stator and further reduces speed.

Reversing

Shaded-pole motors are difficult to reverse because to do so, they must be disassembled. The rotation of the shaded-pole motor is determined by the location of the shaded poles. Figure 7.18 shows a layout of a single-speed, shaded-pole motor. When the shaded poles are on the left side of the main poles, as in Figure 7.18, the rotation will be toward the shaded poles, or clockwise. On the other hand, when the shaded poles are on the right side of the main poles, as shown in Figure 7.19, the rotation will again be toward the shaded poles, but in this case, the rotation will be counterclockwise. Therefore, for reversing of the shaded-pole motor, the stator must be reversed

FIGURE 7.17 Schematic diagram of a three-speed, shaded-pole motor

FIGURE 7.18 Layout of a shaded-pole motor, with clockwise rotation in the direction of the shaded poles

FIGURE 7.19 Layout of a shaded-pole motor, with counterclockwise rotation in the direction of the shaded poles

to change the positions of the shaded poles, and this usually means disassembling the motor.

Troubleshooting

Shaded-pole motors are easy to identify because of the copper band around the shaded pole, previously shown in Figure 7.15. A single-speed, shaded-pole motor is easily diagnosed for trouble because of its simple winding patterns, previously shown schematically in Figure 7.16. Multispeed shaded-pole motors are more difficult to troubleshoot because of the additional speed windings, which were previously shown in Figure 7.17. The shaded-pole motor can be checked with an ohmmeter to determine the condition of the windings. Because a shaded-pole motor has stalled does not mean the windings are faulty. If this condition should occur, the motor probably needs lubrication. The shaded-pole motor is simple and easy to troubleshoot. It is used in many applications in the industry.

7.5 CAPACITORS

The **capacitor** consists of two aluminum plates with an insulator between them. The insulator prevents electrons from flowing from one plate to the other, but it allows storage of electrons. Figure 7.20 shows the schematic symbol for a capacitor. Capacitors are used to boost the starting torque or running efficiency of single-phase motors.

Two Types Used in the Industry

Two types of capacitors are used primarily in the industry: the electrolytic, or starting, capacitor and the oil-filled, or running, capacitor (Figure 7.21). **Starting capacitors** consist of two aluminum electrodes (plates) with a chemically treated paper, impregnated with a nonconductive electrolyte, between them. They can be purchased in ranges from 75 to 600 microfarads (μF) and from 110 to 330 volts. A **microfarad** is the unit of measurement for the strength of a capacitor; all capacitors are rated according to their strength in microfarads. The electrolytic capacitor is used to assist a single-phase motor in starting.

FIGURE 7.20 Symbol for a capacitor

(a) Starting capacitor **(b) Running capacitor**

FIGURE 7.21 Two capacitors used in the industry

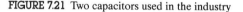

The oil-filled capacitor consists of two aluminum electrodes with paper between them and an oil-filled capacitor case. It is available in microfarad ranges of about 2 to 60 and voltage ranges of 230 to 550. The oil-filled capacitor can be used for small or moderate torque starting, but it is more commonly used to increase a motor's running efficiency.

The major difference between the two types of capacitors is in their purpose. A starting capacitor is built in a relatively small case with a **dielectric**— a nonconductor of electric current. It is used for only a short period of time on each cycle of the motor. Therefore, a starting capacitor has no need to dissipate heat, although its capacity is larger than its counterpart, the running capacitor.

The **running capacitor** is designed to stay in the motor circuit for the entire cycle of operation. Therefore, it must have some means of dissipating the heat. The oil in the capacitor case is used for this purpose. The oil-filled capacitor is physically larger than the starting capacitor but smaller in capacity than the starting capacitor. Both capacitors are in wide use in the industry and serve two basic needs.

Troubleshooting

Short capacitor life and malfunctions can be caused by several different factors. High voltage can cause a capacitor to overheat. This can damage the plates and short the electrodes. Starting capacitors can be damaged by faulty starting apparatus that would keep the capacitor in the line circuit long enough to damage the capacitor. Excessive temperature can shorten the life of capacitors or cause permanent damage due to poor ventilation, starting cycles that are too long, or starting cycles that occur too frequently. The cause of the malfunction should be corrected as soon as possible. The capacitors themselves are frequently the cause of the problem.

All capacitors used on single-phase motors are designed specifically to assist the motor in proper operation. However, in some cases to replace a capacitor with an exact replacement is impossible. If this situation should occur, use the following guidelines for replacing the capacitor:

1. The voltage of any capacitor used for replacement must be equal to or greater than the capacitor being replaced.
2. The strength of the starting capacitor replacement must be at least equal to but not greater than 20% of the capacitor being replaced.
3. The strength of the running capacitor replacement may vary a plus or minus 10% of the capacitor being replaced.
4. If capacitors are installed in parallel, the sum of the capacitors is the total capacitance.
5. The total capacitance of capacitors in series may be found by the following formula:

$$C = \frac{C_1 C_2}{C_1 + C_2}$$

These rules are intended only as a guide. Remember, it is always preferable to use an exact replacement.

Many methods for testing capacitors are in common use in the industry today. A capacitor can be checked by using an ohmmeter. The ohmmeter should be placed on a high-ohm scale and both leads should be connected to the terminals of a discharged capacitor. If the needle of the meter shows a deflection to the right end of the scale and back to zero, the capacitor is good. If the needle comes to rest on 0 ohms, the capacitor is shorted. If the needle of the meter does not move, it indicates an open capacitor.

In case of doubt, another method can be used to check the capacitor. By briefly applying voltage to a capacitor, reading the amperage, and then substituting the values into the formula

$$\text{microfarads} = \frac{2650 \times \text{amperes}}{\text{volts}}$$

we can obtain the exact capacitance. When performing this test, put a fuse in the circuit to prevent overloading due to a shorted capacitor. Starting capacitors should be put in the circuit for approximately five seconds only. Many commercial capacitance testers are available on the market, one of which is shown in Figure 7.22.

7.6 SPLIT-PHASE MOTORS

Two general classifications of split-phase motors are used in the industry. The resistance-start induction run motor and the capacitor-start induction run motor are types of split-phase motors in common use today. Each of these motors has different operating characteristics while being similar in construction. Split-phase motors use some method of splitting the phase of incoming power to produce a second phase of power, giving the motor enough displacement to start. The split-phase motor uses two windings to displace the phase and create the needed displacement between the run and start winding to produce rotation.

Resistance-Start Induction Run Motor

The resistance-start induction run motor shown in Figure 7.23 has both a starting winding to assist the motor in starting and a running winding to continue rotation after the motor has reached a certain speed. Most single-phase motors have some method of beginning the rotation; in a split-phase motor, rotation is started by splitting the phase to make a two-phase current. The single-phase current is split between the running and the starting

FIGURE 7.22 Commercial capacitance tester

FIGURE 7.23 Split-phase motor

winding, which puts one of the windings out of phase by 45 to 90 degrees. The starting windings are used to assist the split-phase motor in starting. They are also used until the motor has reached a speed that is about 75% of its full-capacity speed. The starting windings then drop out of the circuit by use of a centrifugal switch. After that occurs, the motor operates at full speed on the main or running winding alone.

A cutaway view of a capacitor-start-capacitor-run motor is shown in Figure 7.24. Without the capacitors, it closely resembles a split-phase motor.

The split-phase motor can be operated on 110 volts-single phase-60 hertz or 208/230 volts-single phase-60 hertz. Some split-phase motors can operate on either voltage by making simple changes in their wiring if so desired. Thus, it is a dual-voltage motor. Split-phase motors can be reversed by reversing the leads of the starting winding at the terminals in the motor.

Split-phase motors are used when a high starting torque is not required. They are used in such equipment as belt-driven evaporator fan motors, hot-water pumps, small hermetic compressors, grinders, washing machines, dryers, and exhaust fans.

Operation

The phases in a split-phase motor are split by the makeup of the starting windings. The starting winding is designed with smaller wire and more turns than is the running winding, which has a greater inductance. Therefore, the running winding is displaced from the starting winding because of its greater inductance. This displacement causes a resistance to current flow to build up in the running winding. The phase displacement means the current reaches

FIGURE 7.24 Cutaway view of an electric motor

the two windings at separate times, allowing one winding to lead, in this case the starting winding. However, some manufacturers allow current to reach the running winding first by designing an increased resistance into the starting winding and a decreased induction into the running winding. But whatever method is used, the motor basically operates on the same principle: splitting the phase.

The operation of a split-phase motor, referring to Figure 7.25, is as follows:

1. Power is applied to the running and starting windings in parallel. The motor itself splits the phase by the counter electromotive force (emf) in the running winding, which acts as a resistance to hold back the current flow to the running winding. On the alternation of power, the starting winding creates a higher magnetic field than the running winding.

2. In half of a cycle the alternations are changed. The running winding has the stronger magnetic field, moving the rotor a certain distance depending on the number of poles in the motor. For the motor in Figure 7.25, the distance is one-fourth of a rotation.

3. As the alternations continue at the rate of 60 cycles per second, the motor continues to rotate with the magnetic field of the stator. Therefore, the rotor, with its magnetic field permanent, attempts to keep up with the rotating magnetic field of the stator.

4. The motor is equipped with a centrifugal switch that drops the starting

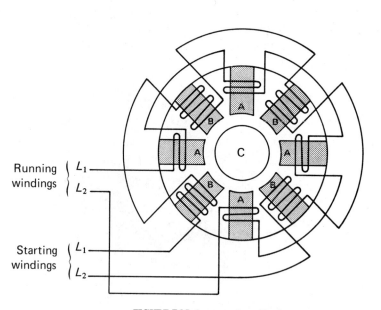

Legend
A: Running windings
B: Starting windings
C: Rotor

Running windings $\begin{cases} L_1 \\ L_2 \end{cases}$

Starting windings $\begin{cases} L_1 \\ L_2 \end{cases}$

FIGURE 7.25 Layout of a split-phase motor

winding out of the circuit when the motor has reached 75% of its full speed.

Troubleshooting

Split-phase motors are one of the most reliable motors used in the heating, cooling, and refrigeration industry. They are used on most types of single-phase equipment. The split-phase motor is easy to troubleshoot if the service mechanic has a good understanding of its operation. The three probable areas of trouble are the bearings, the windings, and the centrifugal switch.

The bearings of any motor often give trouble because of wear and improper maintenance. Identification of a motor with bad bearings is simple. The motor will have trouble turning and in some cases may be locked down completely.

The windings of a split-phase motor can be shorted, open, or grounded. This is easily diagnosed with an ohmmeter.

The centrifugal switch, shown in Figure 7.26, is the hardest section to diagnose for troubles because it stays in the circuit only a short time. The centrifugal switch has a tendency to stick in an open or closed position

FIGURE 7.26 Centrifugal switch used in a split-phase motor

because of wear and often must be replaced. The centrifugal switch can usually be heard when it drops in after the motor is cut off. Hence, it can be checked effectively in this manner. If the centrifugal switch does not drop out of the circuit, the motor will pull an excessively high ampere draw and cut off on overload. One sure method of checking a centrifugal switch is by disassembling the motor and making a visual inspection.

Capacitor-Start Induction Run Motor

The **capacitor-start motor**, shown in Figure 7.27, produces a high starting torque, which is needed for many applications in the industry. The open capacitor-start motor operates like a split-phase motor except that a capacitor is inserted in series with the centrifugal switch and the starting winding. The centrifugal switch breaks the flow of current to the starting capacitor and starting winding. The centrifugal switch opens when the motor has reached a speed that is 75% of its full speed. Figure 7.28 shows a schematic diagram of the motor.

Capacitor-start motors are used on pumps, small hermetic compressors, washing machines, and some types of heavy-duty fans.

Open Type

As we have said, the open capacitor-start motor is similar in design to the split-phase motor with the exception of the capacitor. Therefore, trouble-

FIGURE 7.27 Capacitor-start motor

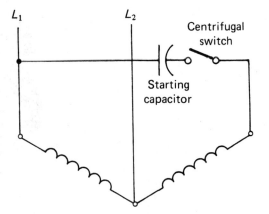

FIGURE 7.28 Schematic diagram of an open capacitor-start motor with a centrifugal switch. Relays may be used instead of a centrifugal switch.

shooting the open capacitor-start motor is similar to checking the split-phase motor except for checking the capacitor. There are four possible areas of trouble in the open capacitor-start motor: windings, bearings, centrifugal switch, and capacitor.

The windings can be easily checked with the use of an ohmmeter by checking for shorts, opens, and grounds.

The bearings of a motor usually fail because of lack of maintenance or wear. Motor bearings will usually become tight or lock down completely. This condition can be determined by trying to turn the motor. If the motor has a tight place in its rotation or will not turn at all, the bearings are faulty in the motor.

Due to the constant opening and closing of the centrifugal switch, it is often the culprit in motor problems. The centrifugal switch may stick in an open or closed position. Or its contacts may be defective. A centrifugal switch, in some cases, can be checked with an ohmmeter to determine its position, open or closed. In other cases, the motor will have to be disassembled to check the switch.

The capacitor is easy to check with an ohmmeter. The capacitor is often mounted in one end bell of the motor rather than on the top of the stator.

Enclosed Type

When capacitor-start motors are used in small hermetic compressors, a centrifugal switch cannot be used because of the oil used to lubricate the compressor. Instead, an external relay is used to break the power going to the starting winding and the starting capacitor. In this case, the capacitor-start motor is an enclosed motor with a starting relay. By inserting a capacitor in the starting winding, a phase displacement is created between the running and starting windings, causing the motor to rotate.

The enclosed capacitor-start motor has an external relay to drop the starting capacitor out of the circuit. This capacitor should be checked to determine its condition.

The condition of the windings of an enclosed motor can easily be checked with an ohmmeter. The windings have a set of terminals on the outside of the casing that lead to the windings. Use an ohmmeter to check across these terminals to determine if the windings are shorted, open, or grounded.

The enclosed motor can also be locked down due to the bearings or to the internal failure of some component of the motor. This condition can be detected with an ammeter or by the humming sound of the motor on an attempted start.

7.7 PERMANENT SPLIT-CAPACITOR MOTORS

Permanent split-capacitor motors (PSC) are simple in design and have a moderate starting torque and a good running efficiency, which makes them a popular motor in the industry. Figure 7.29 shows a PSC motor used as a fan motor. Figure 7.30 shows a hermetic compressor that uses a PSC motor to power the compressor.

The starting winding and the running capacitor of the PSC motor are connected in series, as shown in Figure 7.31(a). The schematic diagram of the hookup is shown in Figure 7.31(b). The running and starting windings are in parallel, but the capacitor causes a phase displacement.

Permanent split-capacitor motors are used on compressors, where the refrigerant equalizes on the "off" cycle, on direct-drive fan motors, and in other applications in the industry. It has a relatively low cost in comparison with other motors because it does not have a switch to drop the starting winding out of the circuit. The PSC motor can be used only when a moderate starting torque is required to begin rotation.

Operation

The permanent split-capacitor motor has two windings: a running (main) winding and a starting (phase) winding. Both windings are wound with almost the same size and length of wire. A running capacitor is put in series with

FIGURE 7.29 Permanent split-capacitor motor used as a fan motor

FIGURE 7.30 Hermetic compressor utilizing a PSC motor

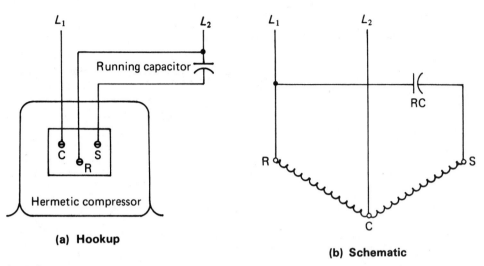

(a) Hookup

(b) Schematic

Legend

C: Common Terminal
R: Running Winding Terminal
S: Starting Winding Terminal
RC: Running Capacitor

FIGURE 7.31 Permanent split-capacitor motor

the starting winding. The capacitor causes the electron flow through the starting winding to shift it out of phase with the running winding. Therefore, a rotating magnetic field is set up, causing the rotor to turn.

Multispeed PSC motors contain additional running windings. The starting winding is in series with the running capacitor and in parallel with the running winding. Figure 7.32 shows a schematic diagram of a three-speed PSC motor. For high-speed operation, the starting and main windings are energized. Medium-speed operation is accomplished by energizing the starting winding with the main and medium-speed windings connected in series. For low-speed operation, the main, medium, and low windings are connected in series with each other, and all are connected in parallel to the starting winding.

Troubleshooting

The PSC motor usually gives trouble-free operation for long periods. The three most common failures in a PSC motor are in the bearings, windings, or capacitor.

The bearings of a PSC motor often become faulty because of wear or lack of proper maintenance. Bearings in any motor can be diagnosed with

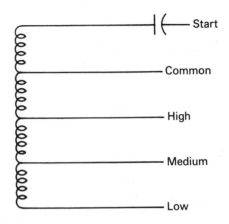

FIGURE 7.32 Schematic diagram of a three-speed PSC motor

little trouble by rotating the motor by hand and noticing rough places in the movement or the shaft being locked in one position.

The windings of a motor become faulty because of overheating, overloading, or a faulty winding. A bad motor winding can be easily checked with an ohmmeter. The windings could be shorted, open, or grounded. The service mechanic should use care in diagnosing problems with the windings of PSC motors because they are often built with several speeds.

A bad capacitor can keep a PSC motor from starting or can pull a high ampere draw when running. Capacitors can be checked by one of the methods covered in Section 7.5. In most cases, faulty PSC motors will be replaced with new motors rather than repaired. The PSC motor is easy to troubleshoot with the right tools and knowledge.

Probably the most difficult aspect of PSC motors is their design. PSC motors are often built with several speeds. Service mechanics must pay careful attention when replacing a faulty PSC motor because if the motor is connected incorrectly, permanent damage can occur. Most PSC motors are furnished with a wiring diagram to ensure correct installation. Motor manufacturers, however, make only a limited number of motors to replace the many different motors in the field. Thus, a service mechanic may have to adapt the replacement motor to a specific application.

7.8 CAPACITOR-START-CAPACITOR-RUN MOTORS

The capacitor-start-capacitor-run motor (CSR) produces a high starting torque and increases the running efficiency. It is actually a capacitor-start motor with a running capacitor added permanently to the starting winding. The starting winding is energized all the time the motor is running. The capacitor-start–capacitor-run motor takes the good running characteristics of a permanent split-capacitor motor (see Section 7.6) and adds the capacitor-start feature. This produces one of the best all-around motors used in the industry.

Capacitor-start–capacitor-run motors are used almost exclusively on hermetic or semihermetic compressors. Rarely will this type of motor be used as an open-type motor because of the cost of the components necessary to produce it. Most open-type motors do not use a starting relay but use the centrifugal switch instead. Open types of motors are usually built as a permanent split-capacitor motor or a capacitor-start motor. Occasionally, a CSR motor will be used in an open-type motor when an extremely high starting torque is required.

Operation

The CSR motor begins operation on a phase displacement between the starting and running windings, which allows rotation to begin. The running capacitor lends a small amount of assistance to the starting of the motor, but its main function is to increase the running efficiency of the motor. Figure 7.33 shows a schematic diagram of this motor.

Troubleshooting

The capacitor-start–capacitor-run motor is sometimes difficult to trouble-shoot because of the number of components that must be added to a regular motor to produce it. The windings, bearings, potential relays, starting capacitor, and running capacitor must all be checked.

Legend

C: Common Terminal
R: Running Winding Terminal
S: Starting Winding Terminal
RC: Running Capacitor
SC: Starting Capacitor
SR: Starting Relay (Potential)

FIGURE 7.33 Schematic diagram of a capacitor-start–capacitor-run motor

The windings of a CSR motor can be easily checked with an ohmmeter to determine if the windings are shorted, open, or grounded. In most cases, the windings will be enclosed in a hermetic casing and the terminals will be on the outside of the casing. However, the type of motor makes little difference in checking the winding as long as the mechanic uses the correct terminals.

The bearings of a CSR motor can be worn so badly that the motor will not turn or will turn only with a great deal of difficulty. The bearings of hermetically sealed motors are enclosed and therefore harder to check, but the conditions of the bearings can be determined by a whining sound, or by the motor pulling a larger-than-normal ampere draw. Care should be taken not to condemn the bearings of a motor because of a high ampere draw unless you are sure this is the problem.

The starting relay can be checked by diagnosing the condition of the contacts and the coil. The contacts can be checked with an ohmmeter or by visual inspection. On an ohmmeter the contacts should show zero resistance. The visual inspection is easy once the relay is disassembled. Then the condition of the contacts can be determined: sticking, pitting, or misalignment. The coil is checked like the windings of a motor.

The starting and running capacitors are easily checked with an ohmmeter to determine their condition.

Troubleshooting a CSR motor is done by checking all components of the motor. These motors must be correctly checked to prevent other components from being destroyed. For example, a capacitor will be destroyed if the contacts or coil of a starting relay are bad.

7.9 THREE-PHASE MOTORS

Three-phase motors are rugged, reliable, and more dependable than other types of motors. The most common type and the type often used in heating, cooling, and refrigeration is the squirrel cage induction type, shown in Figure 7.34. This motor will be the only three-phase motor discussed in this chapter.

Three-phase motors are considerably stronger than single-phase motors because of the three phases that are fed to the motor. Three-phase current actually gives three hot legs to the device, rather than the two hot legs that single-phase supplies. Therefore, instead of having a two-phase displacement, a three-phase displacement is available without using starting components. Three-phase motors are common to the industry; thus, the technician should understand their operation.

Operation

Three-phase motors operate on the same principles as the single-phase

FIGURE 7.34 Three-phase squirrel cage motor

with the exception of the three-phase displacement. A rotating magnetic field is produced in the stator. This interacts and causes a magnetic field in the rotor. However, the three-phase motors require no starting apparatus, because none of the phases are together. In the sine wave of the three-phase motor, none of the phases peaks at the same time. Each phase is approximately 120 electrical degrees out of phase with the others. For this reason, there is no need to use any device to cause a phase displacement, as is needed in the starting of single-phase motors.

Three-phase motors can be purchased in any voltage range desired. For example, a dual-voltage, three-phase motor can be operated on two different voltages with minor modifications in the wiring.

Three-phase motors have two basic types of windings. They are the **star winding**, as shown in Figure 7.35, and the **delta winding**, as shown in Figure 7.36. There is no operational difference between the two types, but it does allow designers more latitude in three-phase-motor design.

Troubleshooting

A three-phase motor can be checked by reading the resistance of the winding with an ohmmeter. If a resistance reading of 0 ohm occurs, the motor is shorted. A reading of infinite resistance indicates an open winding. A reading of some measurable resistance is usually from 1 ohm to 50 ohms, depending on the size of the motor. The larger the motor, the smaller the resistance. The smaller the motor, the larger the resistance of the winding. Care should be taken because of the chance of a spot burnout in the winding. Experience should give service personnel the ability to diagnose any type of electric motor.

FIGURE 7.35 Schematic diagram of the star winding of a three-phase motor

FIGURE 7.36 Schematic diagram of the delta winding of a three-phase motor

7.10 HERMETIC COMPRESSOR MOTORS

Hermetic compressors are becoming increasingly popular because of their low cost. Hermetic motors are of the induction type. They are designed for single- and three-phase current. There are four basic types of single-phase motors used in hermetic compressors. The split-phase is used on small equipment (fractional horsepower). The capacitor-start is also used on small equipment. The permanent split-capacitor is used on most window units and small residential units. The capacitor-start-capacitor-run is used on any application that requires a good starting and running torque. Many hermetic compressors are built with three-phase motors; usually these are used on the larger equipment.

Operation

Hermetic compressor motors are totally enclosed in a shell with refrigerant and oil. Hence, they require special considerations. Nothing can be used inside the shell that is capable of causing a spark or that has to move on the crankshaft, such as a centrifugal switch. Therefore, no starting apparatus can be incorporated inside the compressor shell. Starting relays and capacitors must be mounted and wired outside the motor. It must be remembered that hermetic motors operate the same as other motors with the exception of the enclosure.

Terminal Identification

All single-phase motors have a common, a start, and a run terminal. These terminals are sometimes wired directly into an open-type motor and are difficult to find. The common is the junction point of the start and run terminals. The start and run terminals are connected to one end of the windings while the common is connected to the other end. The schematic diagram of a single-phase compressor motor is shown in Figure 7.37 with the terminals identified. Of course, each of the windings of a three-phase hermetic motor is the same because no starting apparatus is required.

   In single-phase motors, it is important for the service technician to determine the common, start, and run terminals. This task can be performed simply and easily by using an ohmmeter to obtain the resistance of each winding with respect to common. Figure 7.38 shows the ohmic values of a single-phase motor after the resistance has been measured at each terminal on the compressor. To find the run, start, and common terminals, the following procedure should be followed:

1. Find the largest reading between any two terminals. The remaining terminal is common (in Figure 7.38 the reading between *A* and *B* is largest; *C* is common).
2. The largest reading between common and the other two terminals is

Legend

R: Run Terminal
C: Common Terminal
S: Start Terminal

FIGURE 7.37 Schematic diagram of a single-phase compressor with the terminals identified

FIGURE 7.38 Terminals of a single-phase compressor with ohmic values given

start (C to A is 2 ohms and C to B is 10 ohms; therefore, common to B is larger and B is start).

3. The remaining terminal is run (A is run).

This procedure is important, especially in installing the external electric devices, although it is not necessary if a good readable diagram is available. In a three-phase motor, the resistances between all three terminals are the same.

Troubleshooting

Troubleshooting a hermetic compressor motor is often difficult because of its physical makeup and because it is totally enclosed in a shell and cannot be visually inspected. Small hermetic compressors usually have some type of external overload, as shown in Figure 7.39, whereas large hermetic compressors usually have internal overloads, as shown in Figure 7.40. The winding layouts of single-phase hermetic compressors are similar regardless of motor size. The only difference is the size of the windings, which will vary the resistance readings of the motor windings. Three-phase hermetic compressor motors are generally produced in sizes above 3 horsepower. Through experience, the service mechanic will be able to determine the approximate resistance of the motor windings in a hermetic compressor.

FIGURE 7.39 Small hermetic compressor with external overload

FIGURE 7.40 Large hermetic compressor with internal overload (cutaway)

Electrical troubleshooting of hermetic compressor motors is done by taking a resistance reading of the windings with a good ohmmeter. Determining the condition of the windings is easy if the problem with the motor is open windings, shorted windings, or grounded windings. Figure 7.41 shows a schematic representation of these three conditions.

Most single- or three-phase hermetic compressor motors have three terminals on the outside of the casing that connect the motor to the external power wiring, as shown in Figure 7.42. Some large hermetic compressors have more than three terminals, such as dual-voltage, part winding motors or two-speed motors, as shown in Figure 7.43. The resistance readings of single-phase motor windings are not the same because the compressor has a start winding and a run winding connected by a common wire, as shown in Figure 7.44. The physical makeup of a single-phase motor will allow the service mechanic to match the resistance readings to determine the condition of the windings. The sum of the resistance readings of the start to common terminals and the run to common terminals should equal the resistance reading obtained between the run and start terminals, as discussed in the terminal section. If the readings do not match, a spot burnout of the winding is likely. Three-phase motors will have the same resistance in each winding; if not, the motor is bad because of a spot burnout. The service mechanic must be careful, however, before condemning a hermetic motor whose winding resistance readings vary, because the problem may actually be bad connections, a faulty meter, or a misreading of the meter. A good service mechanic should use every possible diagnostic tool to ensure that no good hermetic compressor is condemned.

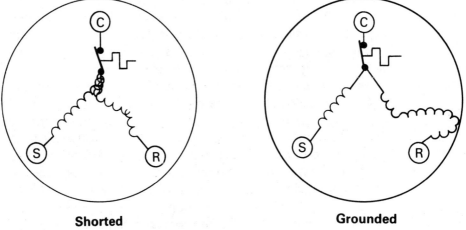

FIGURE 7.41 Schematic representation of good, shorted, open, and grounded compressor windings with internal overload

Diagnosing an open, a shorted, or a grounded hermetic compressor motor is easy because the resistance readings obtained are definite and exact. An open winding in the compressor motor means there is no continuity or no complete circuit; it gives an infinite resistance reading, as shown in Figure 7.45. A shorted winding in a compressor motor means the winding has burned together; it gives a zero ohm reading, as shown in Figure 7.46. A grounded winding in a compressor motor means that part of the winding is contacting the compressor body; it gives a resistance reading between the shell and the

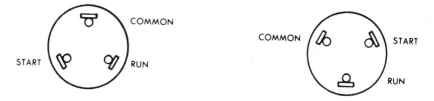

Spade type hermetic push-on terminals

Screw terminals

FIGURE 7.42 Drawings of several terminal arrangements on hermetic compressors

FIGURE 7.43 Drawing of terminals on large hermetic compressor

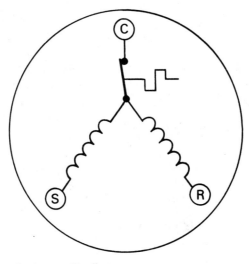

Resistance Readings

C to R: 2 ohms
C to S: 8 ohms
R to S: 10 ohms

FIGURE 7.44 Schematic of windings of single-phase compressor with internal overload (ohm readings for winding shown)

FIGURE 7.45 Drawing of open compressor winding being checked with ohmmeter

FIGURE 7.46 Drawing of shorted compressor winding being checked with ohmmeter

terminals of a compressor, as shown in Figure 7.47. Good contact on the compressor shell must always be maintained if the motor is grounded; therefore any paint must be removed from a small section of the compressor. The open and shorted windings should be read on a low ohm scale (R × 1), but the grounded winding should be read on the R × 10,000 scale or higher. A grounded compressor can be dangerous because the mechanic or consumer can receive an electrical shock if he or she touches the casing of a slightly grounded compressor. A resistance reading as high as 500,000 ohms indicates a grounded compressor that should be changed. Grounded compressors, if allowed to operate, will often operate at a higher than normal temperature: the warmer the windings, the lower the resistance of the ground in most cases.

Before condemning the compressor, service mechanics should make certain that the internal overload of the compressor is not open. This condition can be easily determined by touching the compressor: if the compressor is hot, it is a good indication that the overload is open. The internal overload of a single-phase hermetic compressor is located in the common conductor that connects the run and start windings, as shown in Figure 7.48. Internal overloads used in three-phase hermetic compressor motors are connected at the common junction of the windings, as shown in Figure 7.49. There are many reasons for an internal overload to open in a hermetic compressor:

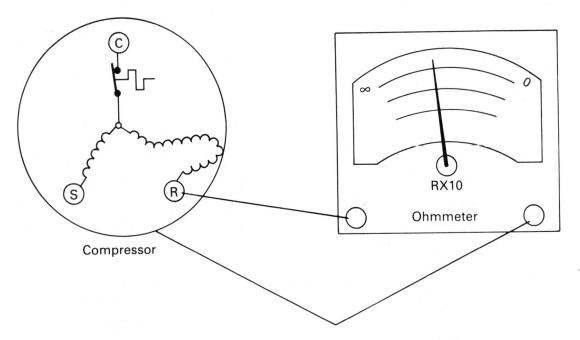

FIGURE 7.47 Drawing of grounded compressor windings being checked with ohmmeter

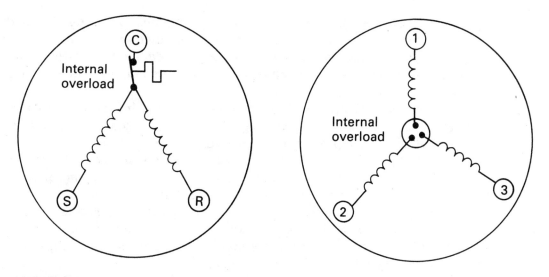

FIGURE 7.48 Schematic of single-phase compressor with internal overload

FIGURE 7.49 Schematic of three-phase compressor with internal overload

for example, low refrigerant charge, locked down compressor, faulty starting components, and high discharge pressure.

Mechanical failures in hermetic compressors often seem like electrical problems, especially when the compressor is locked down or when the internal overload opens because of some mechanical failure. The mechanic must make certain the problem is truly electrical before an accurate diagnosis can be made.

SUMMARY

The single most important operating principle of an electric motor is the rotating magnetic field produced by alternating current. An alternating current is applied to the stator of the motor to produce a magnetic field. The magnetic field interacts with the rotor to produce a magnetic field in it. When these two magnetic fields act together, they produce a rotating movement in the motor.

There are basically five types of motors used in the industry: shaded-pole, split-phase, permanent split-capacitor, capacitor-start-capacitor-run and capacitor-start, and three-phase.

The shaded-pole motor is a low-starting-torque motor used on some propeller types of fans. They are easy to identify because of the copper band around the shaded pole. They are easily diagnosed for trouble because of their simple winding patterns.

Split-phase motors are relatively low torque motors. They are simple and inexpensive devices. A split-phase motor can be used on a small hermetic compressor with the starting winding being dropped out by a starting relay. Or it can be used with any open types of motors that do not require a high starting torque.

The permanent split-capacitor motor has a low starting torque and a running capacitor in the starting winding. The running capacitor remains in the circuit at all times to produce good running efficiency. The permanent split-capacitor motor is used on most residential air conditioners of 5 horsepower or under and on direct-drive fan motors. It is relatively inexpensive because it does not have a switch to drop the starting winding.

The capacitor-start and the capacitor-start-capacitor-run motors are similar in design. They have a high starting torque. The addition of a running capacitor to a capacitor-start motor produces the capacitor-start-capacitor-run motor, which has good running efficiency.

Three-phase motors are commonly used on large pieces of equipment. They operate much like single-phase motors except that they have three basic phase displacements without the use of any starting apparatus. They have better starting and running characteristics than single-phase motors.

Hermetic compressor motors are becoming popular because of their low cost. They are used in many cooling units, especially for the smaller systems. Hermetic motors must have all their starting apparatus wired and mounted externally because it cannot be contained in the compressor shell. A hermetic compressor motor may be any one of the five basic motor types discussed in this chapter. The enclosure distinguishes the hermetic compressor from the other motor types.

Most single-phase motors require some method of producing a second phase of electricity in the motor to make it start. The design of the windings in a split-phase motor allows the use of a centrifugal switch to drop the starting winding out of the circuit after the motor has attained 75% of its full speed. Other types of single-phase motors use capacitors to create the second phase. The permanent split-capacitor motor incorporates a running capacitor to aid in the starting of the motor. The capacitor-start-capacitor-run motor incorporates the running capacitor along with a starting capacitor, using a potential relay to drop out the starting capacitor. The capacitor-start motor operates as a split-phase motor except that a start capacitor is added to the start winding.

The service mechanic will often need to diagnose why a hermetic compressor will not run. Frequently, the fault lies in the hermetic motor. Hermetic compressor motors can easily be diagnosed with a good ohmmeter as open, shorted, or grounded. Spot burnouts are harder to diagnose, but by using the resistance readings of the motor, you can easily do it.

QUESTIONS

1. What is magnetism?

2. What is torque?

3. What is a magnetic field?

4. What is the best conductor of a magnetic field?

5. Unlike poles of magnets _____ each other and like poles _____.

6. What part does polarity play in the operation of an electric motor?

7. How is an electromagnet produced?

8. What part does the frequency of alternating current play in the operation of an electric motor?

9. What is a squirrel cage rotor?

10. What are the five types of single-phase motors used in the industry?

11. List the five types of motors according to their starting strength.

12. Name three examples of the uses of a shaded-pole motor.

13. How does a shaded-pole motor operate?

14. How can a shaded-pole motor be reversed?

15. Draw a diagram of a three-speed, shaded-pole motor.

16. What enables a split-phase motor to develop enough torque to begin rotation?

17. What is the purpose of a centrifugal switch in a split-phase motor?

18. What are the three probable areas of trouble in a split-phase motor?

19. What is the purpose of a capacitor?

20. What is the difference between a running and a starting capacitor?

21. List the five capacitor replacement rules.

22. What is the procedure for checking a capacitor?

23. Explain the operation of a permanent split-capacitor motor.

24. What is the difference between a PSC motor and a capacitor-start-capacitor-run motor?

25. What are the advantages and disadvantages in using the following types of motors? (a) Shaded-pole motor, (b) PSC motor, (c) Split-phase motor, and (d) Capacitor-start-capacitor-run motor.

26. What is the major difference between a split-phase motor and a capacitor-start motor?

27. What are the advantages in using a three-phase motor?

28. Draw a wiring diagram of a capacitor-start-capacitor-run motor.

29. Why must all starting apparatus be mounted externally to a hermetic compressor?

30. What is the process in troubleshooting any electrical motor?

31. The capacitance of an 88 μF and a 108 μF starting capacitor connected in series is which of the following? (a) 196 μF, (b) 96 μF, (c) 48 μF, or (d) 40 μF.

32. The capacitance of two $20\,\mu$F running capacitors connected in parallel is which of the following? (a) $10\,\mu$F, (b) $20\,\mu$F, (c) $30\,\mu$F, or (d) $40\,\mu$F.

33. If a capacitor produces 15 A on 230 V, which of the following is its microfarad rating? (a) $173\,\mu$F, (b) $15\,\mu$F, (c) $200\,\mu$F, or (d) $3450\,\mu$F.

34. You could replace a $35\,\mu$F, 370 V running capacitor with which of the following capacitors? (a) $35\,\mu$F, 330 V, (b) $35\,\mu$F, 390 V, (c) $30\,\mu$F, 440 V, or (d) $40\,\mu$F, 370 V.

35. You could replace a $188\,\mu$F, 250 V starting capacitor with which of the following capacitors? (a) $188\,\mu$F, 110 V, (b) $259\,\mu$F, 110 V, (c) $200\,\mu$F, 250 V, or (d) $300\,\mu$F, 250 V.

36. You could replace a $45\,\mu$F, 370 V running capacitor with which or which combination of the following capacitors? (a) $40\,\mu$F, 440 V, (b) $30\,\mu$F, 250 V, (c) $30\,\mu$F, 370 V, or (d) $15\,\mu$F, 440 V.

37. You could replace an $88\,\mu$F, 250 V starting capacitor with which or which combination of the following capacitors? (a) $180\,\mu$F, 250 V, (b) $180\,\mu$F, 330 V, (c) $150\,\mu$F, 110 V, or (d) $150\,\mu$F, 250 V.

38. Find the common, starting, and running terminals of the following hermetic compressors.

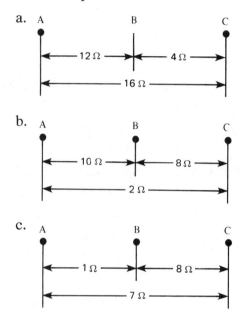

39. Briefly explain the procedure for troubleshooting hermetic compressor motors.

40. What are the failure categories for hermetic compressor motors?

41. What precautions should be taken when checking hermetic compressor motors?

42. What would be the highest allowable resistance readings for a grounded hermetic compressor motor?

8

Components for Electric Motors

INTRODUCTION

In the preceding chapter we discussed electric motors and some of their starting components. In this chapter we will discuss starting components used on single-phase hermetic motors. Single-phase hermetic motors and other special motors require some type of external starting component because they are enclosed in a sealed case. Four types of starting relays are used on this type of motor: current, potential, hot-wire, and solid state. These devices are used on most single-phase hermetic compressor motors with the exception of permanent split-capacitor motors.

Electric motors must have some type of bearings to allow for smooth and easy rotation. The ball bearing or the sleeve bearing are used in most motors. Motors also must have some means of transferring their rotating motion to the device being powered by the motor. A direct-drive hookup transfers the rotating motion directly from the motor to the device. The belt-drive hookup transfers the rotating motion to the device by a belt connection.

Magnetic starters and push buttons are used to stop and start electric motors. A magnetic starter opens and closes sets of contacts to stop and start loads. The magnetic starter also incorporates overload protection for the device it controls. Push-button switches are switches used to control magnetic starters in most cases.

8.1 STARTING RELAYS FOR SINGLE-PHASE MOTORS

Single-phase motors, with the exception of the permanent split-capacitor motor, must have some means of dropping the starting winding (or the starting capacitor in the case of a capacitor-start–capacitor-run motor) out of the circuit. In an open-type motor, this is accomplished simply and easily by a centrifugal switch mounted in the motor. The switch opens the starting circuit once the motor reaches 75% of its full speed. In enclosed motors some type of **starting relay** must be used.

Basically four types of starting relays are used in the industry: current, hot-wire, and solid-state relays are generally used on small hermetic motors. The potential relay can be adapted to any size motor. Starting relays are essential to the operation of a hermetic compressor motor with the exception of the permanent split-capacitor motor.

Each of the four types of starting relays uses a different method to drop the starting circuit in or out. A potential relay operates on the principle of **back electromotive force**. The back electromotive force is the amount of voltage produced in the starting winding of a motor. The current or magnetic type of relay operates on the current or amperage that the motor uses to start. Hot-wire relays use current flow to produce heat across a thermal element, which operates the starting circuit. The solid-state relay uses a positive temperature coefficient (PTC) material that effectively removes the start winding or component from the circuit. Each of these relays must be correctly sized and matched to the application, except the solid-state relays that can be used over certain horsepower ranges. Each relay is designed to remove the starting circuit when the motor reaches approximately 67% to 75% of full speed.

8.2 CURRENT OR AMPERAGE RELAYS

On all electric motors the starting amperage is greater than the running amperage because the rotor is at a standstill on start-up. (The starting amperage of an electric motor is usually stated as the locked rotor amperes [LRA]. The running amperage of an electric motor is usually stated as the full-load amperes [FLA].) Therefore, the ampere draw of the motor is high at the time of the initial start-up. But as the motor gains speed, the ampere draw decreases. The **current** or **magnetic relay** uses the electrical characteristics of the motor to remove the starting circuit electrically once the motor has established a good running speed, approximately 75% of full speed.

Operation

The current relay is built much like a solenoid, with copper wire wrapped around a steel hollow core holding a steel plunger, as shown in Figure 8.1.

FIGURE 8.1 Current relay

Figure 8.2 shows a cutaway view of a current relay. The contacts of the current relay are normally open, as shown schematically in Figure 8.3. The contacts of most current relays are protected by covers that cannot be removed, so it is almost impossible to visually inspect the contacts.

Figure 8.4 shows the position of the contacts when the motor is in the starting phase. As the speed of the motor increases, the amperage decreases. When the motor has reached 67% to 75% of its full speed, the amperage will

FIGURE 8.2 Cutaway of current type starting relay *(Adapted with permission from Texas Instruments, Inc., Attleboro, MA)*

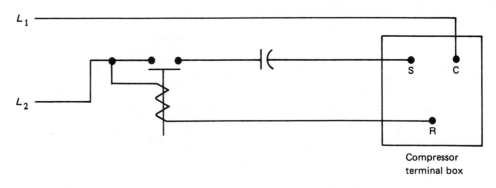

FIGURE 8.3 Schematic of a current relay when it is deenergized

be low enough to cause the magnetic field strength of the relay coil to decrease enough to drop the relay contacts out of the starting circuit.

Troubleshooting

Most current relays are easy to troubleshoot because they have a coil and a set of contacts that can be easily checked with an ohmmeter. The current relay has normally open contacts that are easily checked by turning the relay upside down and checking the contacts with an ohmmeter. If the contacts are good, the relay contacts will read open when the relay is right side up. They will read closed when the relay is inverted. It is imperative when checking a current relay with an ohmmeter that the ohm readings of the contacts be obtained from the correct position on the relay.

The coil of the relay is made of large wire and should have a very low resistance, around 0 to 1 ohm. If the coil reads any higher, more than likely

FIGURE 8.4 Schematic diagram of a current relay when it is energized

FIGURE 8.5 Commercial relay and motor tester

the coil is bad and the relay should be replaced. The current relay and motor starting components should be completely checked to ensure that no other components are bad. Figure 8.5 shows a commercial relay tester that can be used to check current relays.

8.3 POTENTIAL RELAYS

The **potential relay** is gaining popularity because it can be adapted easily to most any compressor. The back electromotive force produced by the starting winding of a motor is the controlling factor of a potential relay.

Operation

When a single-phase motor is operating, a voltage is produced across the starting windings above and beyond the voltage being applied to the motor. The starting windings actually act as a generator to produce the back electromotive force of a motor. The back electromotive force of a motor corresponds to the motor speed. The potential relay is designed to open, dropping the starting circuit, when the motor reaches a certain back electromotive force that is predetermined by the manufacturer of the motor.

Figure 8.6 shows three potential relays that look much like ordinary general-purpose relays. The wiring diagram for a potential relay is shown in Figure 8.7. The potential relay has an advantage over other relays because its contacts are normally closed when the unit starts and causes no arcing. Arcing is an electric spark produced across two sets of contacts. As the motor speed increases, so does the back electromotive force. When the speed approaches 75% to 80% of full speed, the back electromotive force

FIGURE 8.6 Three potential relays

FIGURE 8.7 Schematic diagram of a potential relay

will be large enough to drop the starting circuit by energizing the potential relay coil.

On many permanent split-capacitor hermetic motors that are experiencing difficulty starting, a hard-starting kit can be installed to eliminate the problem, making the PSC motor a capacitor-start-capacitor-run motor. A hard-starting kit consists of a potential relay and a starting capacitor. Figure

FIGURE 8.8 Schematic diagram of a hard-starting kit

(a) Potential relay

(b) Start cap

FIGURE 8.9 Hard-starting kit

8.8 shows the schematic diagram of a hard-starting kit, and Figure 8.9 shows a hard-starting kit as it would come from a manufacturer.

Troubleshooting

The potential relay is easy to troubleshoot because only two parts of the relay must be checked. The coil of the potential relay can be checked with an ohmmeter across 2 and 5 in Figure 8.7, which could check out as good, open, or shorted. The contacts of the potential relay across 1 and 2 should be checked, and they should be closed. Motor current will increase if the potential relay does not drop the starting circuit at the proper time. This

FIGURE 8.10 Commercial potential relay tester

condition is dangerous to the motor and the starting apparatus. The potential relay should always be checked completely to prevent this condition from occurring. A service technician should always check the operation of a unit after the dropout of a starting relay. The correct size relay should always be used for replacement of a faulty relay. If the starting relay is working properly, there will be no ampere draw through the starting circuit after the motor has reached full speed. Figure 8.10 shows a commercial potential relay tester.

The use of potential relays on single-phase hermetic motors has increased over the past 10 years. Potential relays vary greatly in number of terminals, coil group, pickup voltage, and mounting position. An explanation of the numbering system of the two major manufacturers of potential relays should be helpful in the field.

Explanation of G.E. Potential Relay Code

3ARR3– = Potential Relay Type EXAMPLE: 3ARR3-A5C3

1st digit after – = Letter indicating number of term. + bracket

A = 5 screw terminal "L" Bracket. D = 3 screw terminal Flat Bracket.
B = 5 screw terminal Flat Bracket. U = 5 quick connect term.
C = 3 screw terminal "L" Bracket.   "L" Bracket.

2nd digit = Number designating Coil Group

2 = 168 Continuous volts.   6 = 420 Continuous volts.
3 = 332 Continuous volts.   7 = 130 Continuous volts.
4 = 502 Continuous volts.   8 = 214 Continuous volts.
5 = 253 Continuous volts.   10 = 375 Continuous volts.

3rd digit = Letter indicating calibration (Hot Pick Up)

A = 260 – 280 H = 365 – 395 R = 180 – 190
B = 280 – 300 J = 120 – 130 S = 190 – 200
C = 300 – 320 K = 130 – 140 T = 200 – 220
D = 320 – 340 L = 140 – 150 U = 220 – 240
E = 340 – 360 M = 150 – 160 V = 240 – 260
F = 350 – 370 N = 160 – 170 W = 210 – 230
G = 360 – 380 P = 170 – 180

Room Temperature calibration is 5 to 7% lower than these values.

4th digit = Number indicating Mounting Position (See note, end of section)

1 = Face Down
2 = Face Up
3 = Face Out – Numbers Horizontal
4 = Face Out – Rotated 90° clockwise from number 3 position.
5 = Face Out – Numbers upside down – Horizontal.
6 = Face Out – Rotated 90° counterclockwise from Number 3 position.

FIGURE 8.11 Explanation of G.E. potential relay code

Explanation of RBM Potential Relay Code

128– = Potential Relay Type   EXAMPLE: 128-122-1335CA

1st & 2nd digit after – = Numbers indicating Type of Bracket.

    11 = Flat Bracket remote (Tecumseh).
    12 = "L" Bracket (Tecumseh).
    16 = "L" Bracket for "FB" model compressors.
    20 = "L" Bracket for Tecumseh Twins — 1½ HP and Larger.
    21 = "L" Bracket for capacitor box mounting.
    29 = Flat Bracket (Marion) was "14" (under cover).

3rd digit = Contact structure.

    2 = SPNC – less than 1½ HP.
    6 = SPNC – 1½ HP and Larger.

4th & 5th digits = Number of terminals, type and location

    11 = 3 screw terminal.
    12 = 4 screw terminal (seldom used).
    13 = 5 screw terminal.
    23 = 5 quick connect terminals.

6th digit = Number indicating Coil Group

    1 = 130 Continuous voltage   5 = 395 Continuous voltage
    2 = 170 Continuous voltage   6 = 420 Continuous voltage
    3 = 256 Continuous voltage   7 = 495 Continuous voltage
    4 = 336 Continuous voltage

7th digit = Number indicating Mounting Position (See note)

    1 = Face Down
    2 = Face Up
    3 = Face Out – Horizontal – Numbers upside down.
    4 = Face Out – 90° clockwise from No. 3 position.
    5 = Face Out – Horizontal – Numbers right side up.
    6 = Face Out – 90° counterclockwise from Number 3 position.

8th digit = Letter indicating calibration (Hot Pick Up)

    A = 260 – 280 volts.   G = 360 – 380 volts.   P = 170 – 180 volts
    B = 280 – 300 volts.   H = 365 – 395 volts.   R = 180 – 190 volts.
    C = 300 – 320 volts.   J = 120 – 130 volts.   S = 190 – 200 volts.
    D = 320 – 340 volts.   K = 130 – 140 volts.   T = 200 – 220 volts.
    E = 340 – 360 volts.   L = 140 – 150 volts.   U = 220 – 240 volts.
    F = 350 – 370 volts.   M = 150 – 160 volts.   V = 240 – 260 volts.
                                              W = 210 – 230 volts.

Room Temperature calibration is 5 to 7% lower than these values.

9th digit = Letter to indicate customer's part number to be stamped on relay.

NOTE:—Mounting of Relay

As noted above, the 4th digit in the code number of G.E. relays and the 7th digit for RBM relays indicates the position in which the relay is to be mounted. It is of utmost importance that the relay be mounted in the required position. Mounting in any other position can change the relay's operating characteristics enough so that the compressor will not start properly. This can result in burning out the compressor motor.

FIGURE 8.12 Explanation of RBM potential relay code

The two major brands of potential relays are General Electric and RBM. Explanations of their numbering codes are shown in Figures 8.11 and 8.12, respectively. Pay careful attention to the coil group and the pickup voltage when you cross-reference a potential relay.

8.4 HOT-WIRE RELAYS

The **hot-wire relay**, shown in Figure 8.13, operates on the principle that electric energy can be converted to heat. The relay uses two bimetal strips. One strip operates the starting circuit; a second strip acts as an overload for the running winding.

Operation

Figure 8.14 shows the schematic of a hot-wire relay. From L to A is the actual hot wire, which would heat up the bimetal elements B and C. C goes to the starting winding and B to the running winding. If the hot wire reaches a temperature high enough to cause element C to warp, the starting circuit is dropped out. This is the correct procedure. If the hot wire reaches an even higher temperature, this causes element B to warp and open the circuit to the running winding. Then the compressor would cut off on overload by the relay. Some manufacturers allow the hot wire to stretch as it gets hot to open and close the contacts. Both types (warping and stretching) are similar in operation. The schematic in Figure 8.14 shows a hot-wire relay hooked up in a system.

FIGURE 8.13 Hot-wire relay

Legend

L:	Line Voltage Terminal
M:	Main Winding Terminal
S:	Starting Winding Terminal
1,2,3:	Dummy Terminals for Connections
C:	Common Terminal
R:	Running Winding Terminal
S:	Starting Winding Terminal

FIGURE 8.14 Schematic diagram of a hot-wire relay connected to a compressor

Troubleshooting

The hot-wire relay is probably the hardest starting relay to check. The relay is hard to check because it is hard to completely detect the heat being supplied to the relay. Even the slightest temperature difference will cause the relay to show an overload.

The relay contacts are checked with an ohmmeter. The diagnosis of the hot wire and thermal elements of the relay is done while the unit is operating or starting. The first thing to check for is to see if the motor is running. If the motor is running but cutting out on overload, the bearings inside the motor might be tight.

The starting winding should drop in and out quickly and is, therefore, hard to detect. The amperage drawn by the starting winding should be checked to determine if it is too high. Usually, the starting winding will draw a higher amperage than the running winding. As for other starting relays, in troubleshooting the hot-wire relay, the relay and all other motor components should be checked.

8.5 SOLID-STATE STARTING RELAYS

The advancement of solid-state controls over the past five years has produced another type of current-sensitive relay, a PTC starting switch. Certain ceramic materials (PTC) increase their resistance as they heat up from current passing through them. This is the basis of the new solid-state relays that are replacing many of the other varieties on fractional horsepower motors. A PTC solid-state starting relay is shown in Figure 8.15. Its flexibility has greatly expanded its usage because one relay will cover a range of horsepowers. Manufacturers are producing a PTC relay with a built-in starting capacitor, as shown in Figure 8.16, and a built-in starting capacitor and overload, as shown in Figure 8.17.

A new electronic motor starting relay has been introduced that can be used as a universal replacement of many motor starting relays in the industry. This electronic relay operates on a time function principle rather than sensing voltage or current. The time function design allows this relay to be used as a replacement for many potential and current starting relays. This relay is shown in Figure 8.18. The wiring diagram is shown in Figure 8.19 with the relay replacing a potential and in Figure 8.20 replacing a current type relay.

Operation

The solid-state starting relay is placed in series with the start winding of the fractional horsepower hermetic compressor, as shown in Figure 8.21. The

FIGURE 8.15 Solid-state starting relay

FIGURE 8.16 Solid-state relay with built in capacitor

FIGURE 8.17 Solid-state relay with built in capacitor and overload

FIGURE 8.18 Universal motor starting relay *(Courtesy of Sealed Unit Parts Co., Inc., Allenwood, NJ)*

WIRING AS A POTENTIAL RELAY

REPLACES ALL POTENTIAL MOTOR STARTING RELAYS FOR COMPRESSORS RATED AT 110 TO 270 VAC SINGLE PHASE UP TO 5 H.P. ANY PICK-UP VOLTAGE

DIAGRAM NOTE:
A wire from the RUNNING TERMINAL of the compressor motor must be connected to #4 on the relay.

FIGURE 8.19 Wiring of universal relay replacing potential relay *(Courtesy of Sealed Unit Parts Co., Inc., Allenwood, NJ)*

FIGURE 8.20 Wiring of universal relay replacing a current relay *(Courtesy of Sealed Unit Parts Co., Inc., Allenwood, NJ)*

FIGURE 8.21 Schematic of compressor and solid-state starting relay (split-phase motor)

solid-state starting relay normally has a very low resistance. As the compressor motor starts, the current flows to the start winding. The resistance of the solid-state starting relay rapidly rises very high. Therefore, the current flow to the start winding is reduced greatly, which takes the starting winding out of the circuit. Most solid-state starting relays can also be used with a capacitor placed in series with the relay and start winding, as shown in Figure 8.22. One advantage of the solid-state starting relay is that one model covers a variety of horsepower ranges. For example, one relay could be used from 1/12 horsepower to 1/5 horsepower, and another from 1/4 horsepower to 1/3 horsepower. Some manufacturers produce this type of relay to cover compressor motors above 1/3 horsepower. The major advantage of this type of relay is that the service mechanic can stock three basic relays and cover many applications and sizes.

These starting devices are also available with a built-in starting capacitor (refer again to Figure 8.16). When the relay capacitor is used, the existing overload must remain on the compressor for protection. A 3-in-1 relay is also available (refer again to Figure 8.17), which contains the solid-state relay, a capacitor, and an overload.

The solid-state PTC-type starting relay is usually limited to domestic refrigeration applications, but it could be used on other applications if the manufacturer's installation instructions are followed. The solid-state relay usually requires 3 to 10 minutes to cool down between operating cycles, so it is not feasible for short cycling applications.

FIGURE 8.22 Schematic of compressor and solid-state starting relay with capacitor (capacitor-start motor)

Troubleshooting

The solid-state starting relay is easy to troubleshoot because of its simplicity. On many occasions, the relay will show some visible signs that would indicate the condition of the relay. In some cases, the relay will be charred, severely burned, and cracked; in other instances, the relay will only be slightly burned. The relay can also be checked with an ohmmeter: if the resistance is very low, the relay is good; a high resistance across the relay indicates a bad relay. The mechanic must make certain the relay is cool when performing the resistance check. The service mechanic could also check to make certain the relay is dropping the starting winding out of the circuit by using a clamp-on ammeter if a wire is connecting the relay to the compressor terminal. If the relay pushes directly onto the compressor terminals, this check is practically impossible; hence, an amp reading of the compressor motor must be taken, and a judgment must be made if the start winding remained in the circuit.

8.6 MOTOR BEARINGS

All rotating electric equipment have some type of bearing to allow for smooth and easy rotation. A **bearing** is that part of a rotating electric device that allows free movement. The two types of bearings used in the heating, cooling, and refrigeration industry are **ball bearings** and **sleeve bearings**, shown in

Figure 8.23. The ball bearing is the most efficient because it produces less friction. The sleeve bearing is cheaper, however, and is more commonly used.

Ball Bearings

Ball bearings are designed with an inner and an outer ring, which enclose the balls by use of a separator. The inner ring is the bore in which the shaft is pressed.

Ball bearings are lubricated in three ways. The permanently lubricated bearings are sealed at the factory and only rarely require additional lubrication. The packed lubrication of ball bearings is done by hand. The bearing must be disassembled and hand packed with grease every two to five years. Many ball bearings are equipped with a grease fitting and should be lubricated every two years. Lubrication is of prime importance to a motor. And note that overlubrication is as damaging as underlubrication. Figure 8.24 shows a ball bearing pressed into an end bell of a motor.

Ball bearings are used on most heavy loads. However, ball bearings cannot be used in a hermetic compressor because of the danger of sparks. Some of the many advantages of using ball bearings are mountings in any position, antifriction design, versatility, and the ability to carry a large load. Ball bearings are often permanently lubricated from the factory or require infrequent lubrication. Thus, they require less maintenance than a sleeve bearing. The life of a ball bearing is usually long and trouble free, but misuse can rapidly destroy a ball bearing.

FIGURE 8.23 Sleeve and ball bearings used in the industry

FIGURE 8.24 Ball bearing inserted into the end bell of a motor

Sleeve Bearings

Sleeve bearings are brass or bronze cylinders in which a shaft rotates. The bearings have more friction than the ball bearing and thus are used for lighter duty. Lubrication of sleeve bearings is accomplished by oil wick lubrication, yarn-packed lubrication, or oil ring lubrication.

The oil wick lubrication has a wick that extends into an oil reservoir. The wick picks up oil from the reservoir and transfers it to the shaft, as shown in Figure 8.25. The reservoir should be filled with oil twice a year.

The yarn-packed lubrication is merely yarn packed around the shaft to lubricate it, as shown in Figure 8.26. This bearing should be lubricated every few months.

The oil ring lubrication is used on large motors. In this device a ring rotates through an oil reservoir, picking up oil and transferring it to the shaft, as shown in Figure 8.27. The oil level of the reservoir should be checked monthly.

Hermetic compressors use sleeve bearings because with sleeve bearings there is no danger of sparks. Care must be taken in mounting motors equipped with sleeve bearings because of the lubrication problems when the bearings are mounted in a vertical position. Sleeve bearings are commonly used in many applications in the industry.

8.7 MOTOR DRIVES

A motor drive is the connection between an electric motor and a component that requires rotation. Electric motors are used to drive most devices that

FIGURE 8.25 Oil wick lubrication of a sleeve bearing

FIGURE 8.26 Yarn-packed lubrication of a sleeve bearing

FIGURE 8.27 Oil ring lubrication of a sleeve bearing

require rotation. There are two basic driving devices: direct drive and belt and pully drive.

Direct Drive

Direct-drive methods require that a device turn with the same revolutions per minute as the motor. Fan motors and pumps are often direct drive; hermetic compressors are always direct drive. Direct-drive hookups require a close fit between the motor and the device. In a hermetic compressor, the crankshaft is made in one piece, with the motor on one end and the compressor portion on the opposite end. Direct-drive applications usually require a coupling, except with hermetic compressors and fan motors.

The two types of direct-drive couplings used in the industry are the flexible-hose coupling, as used on oil burners, and flange couplings, as used on some open types of compressors and hot-water pumps. Figure 8.28 shows both commonly used types of direct-drive couplings.

V-Belt

Although there are many types of belts used to drive devices, only the **V-belt** will be discussed here. It is used almost exclusively in the industry. By using two pulleys and connecting them with a V-belt, rotation can be transferred from the motor to the device, as shown in Figure 8.29. V-belts are used in many applications in the industry, such as driving open types of compressors, fan motors, and pumps.

There are basically three sizes of V-belts used presently. Size FHP V-belts are ⅜ inch wide and used on fractional-horsepower motors. Size A-section

(a) Flexible hose coupling   **(b) Flange coupling**

FIGURE 8.28 Two widely used direct-drive couplings

is ½ inch wide and used for most jobs requiring 1-to-5-horsepower motors. Size B-section is $2^{1}/_{32}$ inch wide and used when the motor pulley is larger than 5 inches in diameter. More than one V-belt is often used to pull large devices. When belt changes are needed, a matched set should be used in a dual-belt application. The equipment's revolutions per minute can be changed by changing the pulley size or sizes. The following formula can be used:

$$\text{diameter of equipment pulley} = \frac{\text{motor rpm} \times \text{motor pulley size}}{\text{rpm of equipment}}$$

FIGURE 8.29 V-belt connection between motor and fan

(a) Incorrect alignment   **(b) Proper tension and adjustment**

FIGURE 8.30 V-belt alignments

This formula can be transposed to solve for any of the stated values by simple algebra. For example, if the motor rpm were 1750 with a 5-inch pulley, what size equipment pulley would be required to produce a speed of 1000 rpm?

$$\text{diameter of equipment pulley} = \frac{\text{motor rpm} \times \text{motor pulley size}}{\text{rpm of equipment}}$$

$$\text{diameter of equipment pulley} = \frac{1750 \times 5}{1000}$$

diameter of equipment pulley = 8.75 inches

V-belts should always be properly aligned and the tension adjusted. Figure 8.30(a) shows an incorrectly aligned V-belt. The proper amount of tension for most belts is shown in Figure 8.30(b). Frequently a V-belt will slip, causing erratic equipment speed or break. In these cases, the belt should be replaced and the tension adjusted correctly.

SUMMARY

Many types of electric motors require special electrical devices to keep the starting winding or starting component in the circuit until the motor has reached approximately 75% of full speed. Starting relays are used in most cases to drop the start windings and/or starting component from the electrical circuit once the motor has reached operating speed. It is essential the starting components and/or starting windings be removed from the circuit at the correct time to prevent damage to the motor or starting components.

There are four types of relays in common use in the industry today. The current type relay operates on the principle that amperage flowing through a wire will produce a magnetic field. At its highest point, this magnetic field will close a set of contacts, thus putting the necessary starting components in the circuit and removing them at the correct time. The potential relay has a normally closed set of contacts that are opened and closed by the back electromotive force of the motor. This back emf corresponds to the motor speed, thus dropping out the starting components at the right time. Electronic relays used as starting relays are becoming increasingly popular in the industry. They have several advantages over the regular type of starting relay in that they can be used over a variety of sizes while incorporating other motor components if needed. The hot-wire relay is seldom used in the industry because more methods are available from dropping the starting components out of the circuit.

All rotating heating, cooling, and refrigeration equipment must have bearings to operate efficiently and smoothly. Ball bearings and sleeve bearings are commonly used today for this purpose. Correct lubrication is essential to ensure long life and efficient service from the bearings.

In most cases in the industry, electric motors drive some type of equipment. Direct-drive applications are common, as are V-belt applications. Direct-drive applications require very accurate fit with little vibration. The device being driven must turn at the same revolutions per minute (rpm) as the motor. V-belt connections have a certain amount of tolerance, but they must be correctly adjusted. They can be used to alter the revolutions per minute of the equipment by changing the pulleys.

QUESTIONS

1. What is the purpose of starting relays?

2. What is a potential relay and why is it used?

3. What is a current relay and why is it used?

4. What is the purpose of a hot-wire relay? How does it work?

5. The contacts of the current relay are _____.

6. The controlling factor of the potential relay is _____.

7. True or false. The contacts of the potential relay are normally open.

8. What are the two types of bearings used in the industry? What are the advantages and disadvantages of each?

9. What is a direct drive? How is it used in the industry?

10. Why are V-belts popular in the industry?

11. What is the correct tension on a V-belt?

12. What is arcing?

13. True or false. If the starting relay is

working properly, there will be no ampere draw through the starting circuit after the motor reaches full speed.

14. Probably the hardest starting relay to check is the _____.

15. A bearing is the part of a rotating electric device that _____.

16. Name the three ways in which ball bearings may be lubricated.

17. True or false. Overlubrication of a motor and its bearings is as damaging as underlubrication.

18. What is one situation in which ball bearings cannot be used?

19. Name the three ways in which sleeve bearings may be lubricated.

20. Direct-drive hookups require a close fit between _____.

21. What are the common sizes of V-belts and what are their applications?

22. What is the compressor speed of an open-type compressor if the motor rpm is 1,750, the motor pulley diameter is 4 inches, and the flywheel diameter of the compressor is 16 inches?

23. What size motor pulley would be required if the motor turns 1,750 rpm, the fan pulley is 8 inches, and the desired rpm of the fan is 700 rpm?

24. Explain the operation of a solid-state PTC starting relay.

25. What advantages does the solid-state starting relay have over the conventional current-type relay?

LAB SECTION REFERENCE

For experiments and activities dealing with material covered in this chapter, refer to Lab Number 6, Capacitors, and Lab Number 7, Electric Motors.

Contactors, Relays, and Overloads

INTRODUCTION

Control systems used on modern heating, cooling, and refrigeration systems use many different control components to obtain automatic control. The purpose of a control system is to automatically control the temperature of some medium. The function of a control system is to stop and start electric loads that control the temperature of the medium. In the case of an air-conditioning or heating system, the primary purpose is to control the temperature within a certain area. In a refrigeration system, the purpose is to control air temperature or water temperature.

In an air-conditioning or refrigeration system, the compressor is the largest load and usually requires a contactor or magnetic starter to energize it. Loads that require more control and are too large for line voltage thermostats or manual switches use relays for the proper control. Relays and contactors work similarly. The main difference between them is their current-carrying capacity. The contactor can handle a large ampacity. Relays are usually limited in the ampacity they can carry.

In all control systems there must be a means of protecting the loads. The overload is used to protect loads such as compressors, heaters, fan motors, and pumps. The basic overload device is the common fuse. However, the fuse is inadequate to protect effectively all the important loads in a control system. Thus, more effective devices for overload protection are used. As we will see, overloads come in many designs and sizes.

All electric control components serve a definite purpose in the total control system. Fortunately, many manufacturers now use simple control systems with fewer components in their residential air-conditioning control systems. However, the commercial and industrial control systems are fairly complex, with more components and better control than the residential equipment. Therefore, it is essential that heating, cooling, and refrigeration personnel become familiar with control components and understand them so they can diagnose faulty components and perform effective control system troubleshooting.

In this chapter we will look at the components of a control system that control the loads in the system. In succeeding chapters we will discuss other control system components and the methods for troubleshooting these systems.

9.1 CONTACTORS

A contactor (Figure 9.1) is used to control an electric load in a control system. Contactors make or break a set of contacts that controls the voltage applied to some load in cooling systems. They isolate the voltage controlling its magnetic coil from the voltage applied to the load. A contactor consists of a coil that opens and closes a set of contacts due to the magnetic attraction created by the coil when it is energized. Magnetic starters are also used to start and stop large loads in cooling systems. The major difference between magnetic starters and contactors is that the magnetic starter can house its own overload. Magnetic starters will be discussed later in this chapter.

Applications

The largest electric load in any cooling system that requires control is the compressor. In smaller equipment, several other loads might be connected

FIGURE 9.1 Contactors

in parallel with the compressor. Larger systems usually maintain a switching device for each component. The contactor used in a small residential air-conditioning unit probably controls the compressor and condenser fan motor. Figure 9.2 shows a wiring diagram of a small residential unit. Large air-conditioning units usually have several contactors. A large condensing unit, for example, might use two contactors for the compressor and three for the condenser fan motors. Large electric resistance heating systems also have several contactors. For example, they might use a contactor for each section of heaters and for some method of controlling the fan or air supply.

Operation

Different manufacturers design contactors in different ways. But all contactors accomplish the same purpose: opening and closing a set of contacts. The armature of a contactor is the portion that moves. The movement of the armature can be accomplished in basically two ways, with a sliding armature or a swinging armature. The sliding armature is shown in Figure 9.3, and the swinging armature is shown in Figure 9.4. The sliding armature mounts between two slots in the frame of the contactor and moves up and down in these slots. The swinging armature is mounted on a line and moves up and down in a swinging motion.

The armature of a contactor is connected to a set of contacts that causes a completed circuit when the armature is pulled into the magnetic field produced by the coil. This operation is true for both the sliding armature

Legend

C: Contact
COMP: Compressor
CFM: Condenser Fan Motor
HP: High-Pressure Switch
LP: Low-Pressure Switch

FIGURE 9.2 Schematic diagram of a small residential air-cooled condensing unit with a contactor controlling the compressor and condenser fan motor

Sliding
Armature

Swinging
Armature

FIGURE 9.3 Contactor with a sliding armature   **FIGURE 9.4** Contactor with a swinging armature

and the swinging armature. The magnetic field that closes a contactor is created by a coil wound around a laminated iron core, as shown in Figure 9.5. When the coil is energized, a magnetic field is created around the laminated core. The core then becomes an electromagnet of sufficient strength to attract the armature closing the contacts. Both types of contactors use the same principle of operation. Some contactors have springs mounted between the armature and the stationary contacts to ensure the contactor opens when the coil is deenergized.

Coils

Coil characteristics depend on the type of wire and the manner in which it is wound. The potential coil is energized by a certain voltage being applied to

FIGURE 9.5 Solenoid coil of a contactor

it. Coils of this type are designed to be operated on 24 volts, 110 volts, 208/230 volts, and occasionally 460 volts. The coil is identified by the voltage marked on it (refer again to Figure 9.5). The potential coil is used on many special relays in the industry. The connection of a coil is usually made directly on the terminals of the coil. But in some cases, the connections are jumped to a section of the contactor frame.

Contacts

The contacts of a contactor make a complete circuit when the contactor is energized, allowing voltage to flow to the controlled load. Contactors are rated by the ampere draw they can carry. There are two types of loads that a contactor can control: an inductive load, such as a motor, which has a higher ampere draw on start-up than while running; and a resistive load, which has a constant ampere draw, such as a resistance heater. Some contactors are rated for both inductive loads and resistive loads. So care should be taken when selecting a replacement contactor. The ratings of contactors are usually marked on the contactor frame.

Contacts are made of silver and cadmium, which resists sticking. The contacts are connected to a strong backing by mechanical or chemical bonding. The chemical composition of contacts is such that they operate at cool temperatures at up to 125% of their current-carrying capacity. Contactors are usually manufactured with two or three poles and in many cases four. The fourth pole is used to interlock some load device into the system or can be left unused. A two-pole contactor is required for single-phase systems. A contactor with at least three poles is required for three-phase systems.

Troubleshooting

The diagnosis of a faulty contactor encompasses three sections of the contactor: the coil, the contacts, and the mechanical linkage. A defect in any contactor part can cause the total contactor to be faulty.

Coil. The contactor coil must be in good condition to create a strong enough electromagnetic force to pull in the contactor. The coil of a contactor rarely becomes so weak that it does not close the contacts, unless there is excessive friction of the mechanical linkage. The coil can be diagnosed as good, open, or shorted. The open and shorted conditions indicate a bad coil and can be checked with an ohmmeter. If the coil is shorted, the resistance reading will be zero. If the coil is open, the resistance will be infinite. A measurable resistance usually indicates a good coil.

A coil can also be checked by applying voltage to it and observing the contactor to see if it closes. The voltage reading of a coil should be taken before checking the coil to see if the contactor should be closed. Care should be taken when the diagnosis leads to a shorted contactor coil. If voltage is applied to it, the coil will cause a direct short and other damage could result.

Contacts.   The contactor contacts must be in good condition to ensure that the proper voltage reaches the load. In most cases, a visual inspection is sufficient to diagnose bad contacts. Figure 9.6 shows a good set and bad set of contacts for comparison.

In some cases, a voltage reading taken across the contacts of the same pole will show the voltage drop across the contacts. Figure 9.7 shows the proper procedure for testing a set of contacts in this manner. The voltage indicated on the meter is the voltage drop across the contacts (the voltage lost to the equipment). The 20 volts shown on the meter are considered to be excessive. Hence, the contactor should be replaced or repaired. Any voltage above 5% of the rated voltage for the equipment is considered to be excessive. The contactor must be closed with voltage applied to make this check.

Mechanical Linkage.   Probably the easiest faults to diagnose with a contactor are problems with the mechanical linkage. In most cases, any trouble with the mechanical linkage can be detected by visual inspection. Or problems can be detected by breaking the power supply and manually moving the armature of the contactor to see if the movement is free and without excessive friction. The mechanical linkage of a contactor will usually fail because of wear, corrosion, or moisture. In many cases when a contactor coil burns out, it will heat the coil and cause the varnish of the coil to gum up the contactor. For a contactor to operate properly, it must seat the contacts accurately and have free moving parts.

(a) Good set   **(b) Bad set**

FIGURE 9.6 A good set and a bad set of contacts

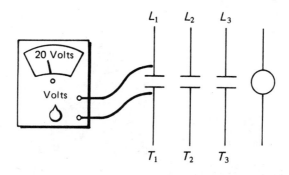

FIGURE 9.7 Procedure to use in testing a set of contacts with a voltmeter

Repairing. Contactors can be repaired by using replacement parts from the manufacturer or a wholesaler if time permits. However, it is often difficult to find all the necessary parts, such as contacts and coils, because it is almost impossible for a wholesaler to stock all the components needed for all contactors. Most manufacturers do sell a kit that will completely replace the contact portion of the contactor. But since parts are difficult to obtain, it is usually advisable to purchase a new contactor instead of repairing the faulty one. However, be careful to choose the correct contactor for the particular application and size.

9.2 RELAYS

Relays are used to open and close a circuit to allow the automatic control of a device or circuit. Relays are similar to contactors with the exceptions of the pole configuration and the amount of current that each device can effectively handle. Relays can be used to control most any device in the system within a certain ampacity limit.

Operation

Relays are built with the same components as a contactor. There are a coil, contacts, and some type of mechanical linkage to open and close the contacts when the relay coil is energized. When voltage (or current in some cases) is applied to a relay, it will close because of the magnetic field created in the coil and iron core. This magnetic field causes the armature of the relay to be attracted to the electromagnet created by the coil and its core. Figure 9.8 shows several different commonly used relays. The coil of a relay can be energized by voltage or current draw. The general type of relay controlling a device is closed by voltage. The current relay is used to

FIGURE 9.8 Several different types of relays used in the industry

control some starting device when used with a small hermetic motor.

Relays can be purchased with most any type of pole configuration. Normally open or normally closed contacts are both used in control circuits. The normally open contact opens when the relay is deenergized and closes when the relay is energized. The normally closed contact closes when the relay is deenergized and opens when the relay is energized. The normal position of the relay denotes the position of any controlling device in the deenergized position. The most common types of pole configuration for relays are single-pole–single-throw, single-pole–double-throw, double-pole–single-throw, and double-pole–double-throw. Any of these configurations can be normally open or normally closed.

Applications

Relays can be used to control indoor fan motors, condenser fan motors, damper motors, starting capacitors, and control lockouts. They are used for any device that requires an automatic means of opening and closing a circuit.

The indoor fan relay is a good example of the use of a relay. On the cooling cycle, the indoor fan must be energized. This is accomplished by use of an indoor fan relay, as shown in the diagram in Figure 9.9. The indoor fan relay will energize when the system switch and the fan switch or the

L_1

L_2

Legend

IFM: Indoor Fan Motor

IFR: Indoor Fan Relay

G: Fan Terminal on Thermostat

Y: Cool Terminal on Thermostat

R: Voltage Terminal on Thermostat

FIGURE 9.9 Schematic diagram of an indoor fan relay circuit

cooling thermostat are closed, starting the fan motor. The indoor fan relay can also be used to control a two-speed fan motor, using high speed on cooling and low speed on heating, with a thermal fan switch controlling the fan motor on heating, as shown in Figure 9.10. Relays can also be used to control large contactors that cannot be effectively energized with 24 volts in a control system, as shown in Figure 9.11.

The potential or voltage-type relay is energized when voltage is applied to the relay coil. This relay is used to control some load device by opening and closing its contacts. The voltage-type relay can be used for many purposes, such as indoor fan relays, condenser fan relays, control relays, and lockout relays.

Three types of relays are used to assist in starting motors, as we saw in Chapter 8. The potential relay uses voltage to energize its coil and drop the starting apparatus out of the circuit. The current relay uses current flow to energize the circuit that contains the starting apparatus and then drops the circuit out when the current has dropped. The thermal relay uses heat to open and close starting circuits when it is used with a motor. Starting relays for motors are used throughout the industry. However, these relays should not be confused with the general type of relay used to control loads.

Construction

The contacts used in relays are made like contacts in a contactor. The contact is made of a silver and cadmium alloy attached to some kind of strong backing that can withstand the pressure exerted by the armature.

A relay is usually mounted in a plastic enclosure. Hence, visual inspection is not as easy for relays as it is for contactors. The relay contacts cannot be seen unless the relay is disassembled or the cover is removed. The contacts of a relay are shown in Figure 9.12.

Legend:

IFR: Indoor Fan Relay
IFM: Indoor Fan Motor
C: Contactor
FS: Fan Switch

FIGURE 9.10 Schematic diagram of indoor fan relay on heating/cooling application

The relay armature can be swinging or sliding as shown in Figure 9.13. These devices operate in relays in the same way that they do in contactors.

The relay coil is built to produce enough magnetism to effectively close the contacts of the relay. The size of the relay coil is smaller than the con-

Legend:

COMP:	Compressor
C:	Contactor
IFR:	Indoor Fan Relay
IFM:	Indoor Fan Motor
CR:	Control Relay
HPS:	High-pressure Switch
LPS:	Low-pressure Switch
CR:	Control Relay
CH:	Crankcase Heater
TRANS:	Transformer
CIT:	Compressor Internal Thermostat

FIGURE 9.11 Schematic diagram of control relay used to energize contactor

tactor coil. Figure 9.14 shows a comparison of a coil used in a relay and a coil used in a contactor.

Troubleshooting

The diagnosis of a faulty relay is done in much the same way as the diagnosis of a faulty contactor. Diagnosis of a coil is the same whether it is for a

FIGURE 9.12 Contacts of a relay

(a) Swinging armature

(b) Sliding armature

FIGURE 9.13 Two types of relays used in the industry

(a) Contactor   (b) Relay

FIGURE 9.14 Comparison of solenoid coils for a contactor and a relay

relay or for a contactor. The relay contacts are usually hidden and cannot be visually inspected without disassembly. They must be checked with an ohmmeter. The contacts of a relay are not as heavy as the contacts of a contactor and therefore can take less punishment. Normally open and normally closed relay contacts will be completely melted if a different phase of voltage is applied. This must be taken into consideration when troubleshooting contacts of relays.

The mechanical linkage of a relay gives less trouble than a contactor because of the lighter weight of the armature. Any mechanical linkage problem in a relay will usually be caused by sticking contacts.

9.3 OVERLOADS

An overload is an electric device that protects a load from a high ampere draw by breaking a set of contacts. The simplest form of overload protection is the fuse. Fuses can be used to protect wires and noninductive loads, but they provide inadequate protection for inductive loads. A load that is purely resistive in nature with no coils to cause induction is called a **noninductive load**. The most common noninductive load used in the industry is an electric heater.

Fuses

Fuses consist of two ends or conductors with a piece of wire that will melt and break the circuit if the current passing through it exceeds the amperage rating of the fuse. Fuses are available in many different styles and designs (again see Chapter 4). Fuses are used most commonly to protect wires, circuit components, and noninductive loads. Electric resistance heaters are the most common noninductive loads protected by fuses. Figure 9.15 shows a schematic diagram of a set of resistance heaters protected by fuses. The system in Figure 9.15 controls three electric heaters by energizing contactors to start the heaters. The fuses in the circuit are used as safety devices for the heaters.

Circuit breakers are used for the same purpose as fuses but allow a high starting load. However, many control circuits use fuses for protection.

Line Break and Pilot Duty Overloads

Overload devices used to protect inductive loads are more effective devices than fuses but also are more complex. Inductive loads require more amperage to start than to run. The amperage of a motor at the moment power is applied is largest because the rotor of the motor is in a stationary position. Figure 9.16 shows a graph of the ampere draw of a motor from start to full speed.

Overloads can be divided into two basic groups: line break and pilot duty. The **line break overload** breaks the power to a motor. A **pilot duty overload**

NOTE: Fuses are used as overloads for H1, H2, and H3

Legend

HC1: Heater Contactor 1   FU1: Heater 1 Fuse   H1: Heater 1
HC2: Heater Contactor 2   FU2: Heater 2 Fuse   H2: Heater 2
HC3: Heater Contactor 3   FU3: Heater 3 Fuse   H3: Heater 3

FIGURE 9.15 Schematic diagram showing fuses used as overloads for protection of resistance heaters

FIGURE 9.16 Ampere draw for motor speeds from locked rotor to full speed

breaks an auxiliary set of contacts connected in the control circuit. Overloads can be manual or automatic reset; the manual reset must be reset by the service mechanic or consumer, but the automatic reset will reset automatically.

Line Break Overload. A line break overload is shown in Figure 9.17. One of the most common types of line voltage overload is a metal disc mounted between two contacts. It is called a bimetal line break overload. Figure 9.18(a) shows a schematic diagram of a bimetal overload in closed position. If the current draw or temperature of the motor is sufficient to cause the disc to overheat and expand, the contact would open, as shown in Figure 9.18(b). This breaks the flow of power to the load. In some overloads, a heater or wire installed below the disc is sized to give off heat, as shown in Figure 9.19. This gives a more accurate range of protection. Figure 9.20 shows a fractional horsepower compressor with an in-line overload.

FIGURE 9.17 Line break overload

(a) Closed   **(b) Open**

FIGURE 9.18 Schematic diagram of bimetal line break overload in closed and open positions

FIGURE 9.19 Bimetal overload with a heater installed below the disc

FIGURE 9.20 In-line overload on fractional horse-power compressor

Another type of bimetal overload is the three-wire klixon overload. It uses the current draw of both windings to open or close the overload and break the common to both windings, as shown schematically in Figure 9.21.

The most popular line break overload for use in small central residential systems is an **internal compressor overload**, shown in Figure 9.22. The internal overload is a small device inserted into the motor windings. This overload can sense the current draw of the motor as well as the winding temperature more effectively than external overloads. Figure 9.23(a) shows a schematic diagram of a compressor with a bimetal internal overload. Figure 9.23(b) shows a compressor with a three-wire bimetal overload. Some small three-phase hermetic motors also use an internal type of overload. Internal overloads should not be confused with internal thermostats. They are similar in appearance. The purpose of an internal thermostat is to break the control line if the windings overheat.

FIGURE 9.21 Schematic diagram of a three-wire bimetal overload

FIGURE 9.22 Internal compressor overload

(a) Bimetal overload (Internal) (b) Three-wire bimetal overload (External)

FIGURE 9.23 Schematic diagram of bimetal overloads

Pilot Duty Overload. The pilot duty overload breaks the control circuit when an overload occurs, which would cause a contactor to be deenergized, as shown in Figure 9.24. This type of overload is common on larger systems and still exists on smaller systems presently in the field.

Two basic pilot duty overloads are being used in the industry today: the current overload and the magnetic overload. The **current overload** is shown in Figure 9.25. This overload works similarly to the line break overloads except that a pilot duty set of contacts is opened rather than the line. In most cases, the bimetal disc of the overload would have to be so heavy that it could not control line voltage effectively. Therefore, in larger overloads pilot duty contacts are used.

The **magnetic**, or Heinemann, **overload**, as shown in Figure 9.26, is another

FIGURE 9.24 Schematic diagram showing pilot duty overloads in the circuit

FIGURE 9.25 Current-type pilot duty overload

FIGURE 9.26 Magnetic overload

type of pilot duty overload used in the industry. The magnetic overload consists of a movable metal core in a tube filled with silicone or oil. Surrounding the metal tube is a coil of wire. When the current increases, so does the magnetic field of the coil. The overload operates by the magnetic field created by the coil. The device is designed to create a magnetic field that is strong enough to pull the core up, opening the pilot contacts on overload.

The magnetic overload has a time-delay feature. There is a small hole drilled in the core. Once the field begins pulling the core in, the oil or silicone must go from one end of the tube to the other through the small hole. Thus, there is a short interval, due to the oil flow, between the time the motor starts up and the time the overload would break the circuit. Figure 9.27 shows a schematic diagram of magnetic overloads in a compressor circuit.

Many electronic solid-state overloads are used in the industry, and more are being developed. The design of the electronic overload varies with the application, but in most cases a temperature sensor is placed in the motor winding and connected to an external module mounted in the control panel of the unit. The sensors mounted in the motor windings will change resistance with a change in motor winding temperature. This change in resistance, when amplified by the electronic module overload, will open a set of pilot duty contacts. Most electronic overloads use two or three sensors mounted in the motor windings. This method of overload protection gives rapid response time to an overload condition. An electronic overload with a sensor is shown in Figure 9.28. A more complete explanation of the electronic overload is given in Chapter 14.

Troubleshooting

Most overloads are easy to troubleshoot because they have only one set of contacts along with a heater to heat the bimetal disc. An ohmmeter can be used to test the contacts in both pilot duty and line break overloads. In some cases, there will be a break in the heater or coil, depending on the type

Magnetic overloads

FIGURE 9.27 Schematic diagram showing magnetic overloads protecting a compressor

FIGURE 9.28 Electronic overload with sensor

of overload you are checking. This element of an overload can also be checked with an ohmmeter.

The internal overload of a compressor is more difficult to diagnose because it cannot be isolated from the system. Most internal overloads break the common of the motor. If the motor is checked and there is an open starting and running winding, chances are that the common or the internal overload is open. If the compressor housing is cool, the motor might be damaged rather than the overload. Be sure when making this test that the compressor is cool. If the compressor is warm or hot, the overload could be open and working properly.

9.4 MAGNETIC STARTERS

A **magnetic starter**, shown in Figure 9.29, is composed of four sets of contacts, a magnetic coil to close the contacts when the coil is energized, and a set of overloads. Figure 9.30 shows a magnetic starter with the parts labeled. The coil is the heart of the system. All functions of the total control, and stopping and starting the equipment, are accomplished by the coil being energized or deenergized. The contacts open and close, depending on the action of the coil. The magnetic starter also has a means of overload protection, which deenergizes the coil in the event that an overload occurs. Figure 9.31 shows a simple wiring diagram of a starter controlled by a thermostat.

In the heating, cooling, and refrigeration industry, many of the motors used are protected by internal overloads in the motor or by some external means, such as an overload relay, magnetic overload, or thermal elements. A contactor is a device that opens and closes automatically and allows voltage to flow to the equipment (refer again to Chapter 4). Most air-conditioning equipment uses a contactor with a separate means of overload protection. In many cases, especially with three-phase equipment, manufacturers use a magnetic starter, which incorporates a means of overload protection along with the ability to stop and start current flow to the equipment.

FIGURE 9.30 Parts of a magnetic starter

FIGURE 9.29 Magnetic starter

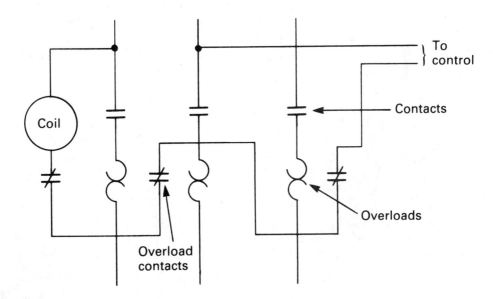

FIGURE 9.31 Schematic diagram of a magnetic starter

Types of Magnetic Starters and Their Operation

The overloads used in a magnetic starter are of three general classes: bimetal relay, thermal relay, and molten-alloy relay. The **bimetal relay** is composed of two metals welded together with different expansion qualities. One end is fastened securely and the opposite end can move. When this element is heated above a certain temperature due to the current flow to the equipment, it will break, causing the starter to open.

The **thermal relay** operates on the heat going through a wire to determine the current flow. If the heat is great enough to indicate an overload, the contacts of the relay shown in Figure 9.32 will open.

The molten-alloy overload element, shown in Figure 9.33, is nothing more than a link of material that is a good electrical conductor. A ratchet wheel is soldered to the conductor with a special alloy. When the current flow produces a higher-than-normal temperature, the solder melts, allowing the ratchet wheel to open the contacts.

No matter what type of overload relay is used, it should be sized correctly to work properly. Manufacturers of magnetic starters distribute much information on how to size starters and overloads (sometimes referred to as "heaters").

Troubleshooting

Magnetic starters are used on most three-phase equipment because they contain overloads and can effectively control the operation of loads in a heating, cooling, or refrigeration system. Some common faults of magnetic starters occur in the contacts, solenoid coil heaters or overloads, and the mechanical linkage. No matter what part of the magnetic starter is faulty, it

FIGURE 9.32 Thermal relay

FIGURE 9.33 Molten-alloy overload element

must be treated as a unit at the beginning of the diagnosis.

The service mechanic must first be certain that the magnetic starter is at fault. If the magnetic starter is at fault, it must be replaced, or the faulty section must be repaired. The contacts of a starter can be checked by visual inspection or by checking the voltage drop across each set of contacts. If the contacts are badly pitted, or if there is a voltage drop across the contacts, then they must be replaced.

The coil of a contactor can easily be faulty; it can be shorted, open, or grounded. The linkage that connects the movable contacts to the solenoid plunger can be worn or broken and will not close the contacts even if the coil is energized.

The overloads are difficult to troubleshoot because of the different types. Overloads are designed to open a control circuit when an overload exists. The overload circuit of a magnetic starter is connected within the control enclosure. The service mechanic must be familiar with the types of overloads and their operation. The size of the overload should also be checked. The mechanic can easily check a magnetic starter step by step after it has been determined that the starter is not closing and should be.

9.5 PUSH-BUTTON STATIONS

Push-button stations or switches are switches that are controlled manually by the pressing of a button. Figure 9.34 shows a start-stop push-button station. Figure 9.35 shows two wiring diagrams using push-button switches as controls. They can have two switches (for stop and start) or many switches (for such functions as on, off, start, stop, jog, reverse, and forward). They are designed in many forms and with many different functions. There are relatively few instances in the heating, cooling, and refrigeration industry where push-button stations are used. They are used mainly for motor control, circuit control, and magnetic starters.

Push-button stations are easy to troubleshoot. The service mechanic must know, or be able to find out, the normal position of the switch. In some cases, push-button switches may be quite complex, but this will not occur very often. An ohmmeter can be used to diagnose the condition of the switch.

SUMMARY

Control systems used on modern heating, cooling, and refrigeration systems use many different control components to obtain automatic control. In this chapter we described four of these components: contactors, relays, overloads, and magnetic starters.

FIGURE 9.34 Stop-start push-button station

(a) Stop-start station

(b) Stop-forward-reverse station

FIGURE 9.35 Wiring diagrams of some common push-button control systems

Contactors play an important part in the correct operation of equipment in the industry. Their purpose is to make and break a power circuit to a load, thus allowing the proper control of the equipment due to the contactor being energized or deenergized. The contacts, coil, and mechanical linkage are the main parts of the contactor. Each part must work properly for the contactor to work correctly. The coil can be faulty by being shorted or open. If it is good, a measurable resistance can be obtained on an ohmmeter. The contacts can be faulty due to wear, corrosion, or excessive friction. Whatever the diagnosis of a contactor, it should be repaired or replaced whenever it is faulty or gives erratic operation. When replacement is necessary, it is essential to select the correct contactor for the job.

The relay is used to control many loads in control systems. It is the most widely used device in the industry. The application and size of relays should be identified before any replacement is attempted. The relay has the same components as a contactor: coil, contacts, and mechanical linkage. The coil is easily diagnosed with an ohmmeter as good, open, or shorted. The contacts must be in good shape for proper operation and can be checked with an ohmmeter or voltmeter. The contacts of a relay are usually enclosed, which prevents a visual inspection. A faulty mechanical linkage in a relay usually results in sticking contacts.

Overloads play an important part in the industry because they protect expensive loads. The fuse is the simplest type of overload protection used. It is effective on noninductive loads, in protecting wires, and in protecting circuit components.

Overloads are divided into two basic types: line break and pilot duty. The line break overload breaks the line voltage to the components. It is used on small hermetic compressors and motors and is connected directly in the line voltage supply to the equipment or load. The pilot duty overload breaks a pilot duty set of contacts in the control circuit. Pilot duty overloads are arranged so that the line voltage feeds directly through them and on to the load. This line voltage will create an operating condition in the element in the overload—thermal, current, or magnetic—and open a set of pilot duty contacts. The internal overload is mounted directly in the windings of the motor and usually has no connections to the outside of the motor. The internal overload is connected in series with the common of the motor. All overloads must be properly sized to do an adequate job of protection.

A magnetic starter is much like a contactor; it has contacts that are opened or closed depending upon the action of the solenoid coil that energizes or deenergizes the magnetic starter controlling a load. The magnetic starter usually has three or four sets of contacts. It provides overload protection to the load it controls. The overload in the magnetic starter must be sized to the load it is controlling. The overload contacts remain in the starter and the overload elements, called heaters, must be correctly sized for the load. The heaters are installed separately in the starter. Push buttons and stations are used with magnetic starters in many applications but usually not in air-conditioning equipment. The most common type of push button is a simple start-stop station.

QUESTIONS

1. What is the largest electric load in an air-conditioning or refrigeration system?

2. What is the major difference between a relay and a contactor?

3. The major difference between magnetic starters and contactors is that the magnetic starter _____.

4. How does a contactor operate?

5. What are the two types of armatures used in contactors? How does each operate?

6. Name the three major sections of a contactor.

7. What is the proper procedure for checking the coil of a contactor?

8. Contacts are made of _____.

9. What are the major reasons for replacing the contacts of a contactor?

10. True or false. A voltage reading taken

across the contacts of the same pole will show the voltage drop across the contacts.

11. The easiest faults to diagnose with a contactor are problems with _____.

12. Why is it difficult to repair contactors?

13. Name the components of a relay.

14. How does a relay operate?

15. What do the terms "normally open" and "normally closed" mean in reference to relays and contactors?

16. True or false. The contacts in a relay are easily checked by a visual inspection.

17. True or false. The size of the relay coil is smaller than the coil of a contactor.

18. What is the purpose of an overload?

19. What is a resistive load?

20. What is an inductive load?

21. Fuses are used most often to protect _____.

22. The line break overload breaks _____.

23. How does a line break overload operate?

24. True or false. The most popular line break overload for use in small equipment is the three-wire klixon overload.

25. The pilot duty overload breaks _____.

26. Name the two basic types of pilot duty overloads and describe how they operate.

27. What is the advantage of an internal compressor overload?

28. What is the proper procedure for checking an internal compressor overload?

29. What is a magnetic starter?

30. What are the three types of overloads used in magnetic starters in the industry? How does each type work?

31. True or false. Magnetic starters are primarily used on single-phase equipment.

32. True or false. Push-button stations are widely used in the industry.

LAB SECTION REFERENCE

For experiments and activities dealing with material covered in this chapter, refer to Lab Number 9, Contactors and Relays, and Lab Number 12, Overloads.

10

Thermostats, Pressure Switches, and Other Electric Control Devices

INTRODUCTION

In the preceding chapter we discussed some commonly used devices that control loads in heating, cooling, and refrigeration systems. In this chapter we will discuss devices that control some small loads and contactors and relays that control larger loads.

Thermostats are extremely important to the industry because they are opened and closed by a change in temperature. In all phases of the industry, we must control temperature and in most cases we do this by use of a thermostat. Pressure switches are sometimes used to control temperature by the pressure-temperature relationship, but in most cases pressure switches are used as safety devices. Transformers are used to reduce line voltage to the low voltage that is used in many control systems.

Thermostats play an important part in most control systems. The thermostat is commonly used as the primary control to control the temperature within a given area. In some cases, it is also used as a safety device, as in motor protection. Thermostats that are used as primary controls can be heating thermostats or cooling thermostats or some combination of the two. Thermostats can also be used for a staging effect, that is, for operating equipment at different times, depending on the demands put on the system. Staging can provide a main stage for normal operation and a secondary stage for energizing a specific part of the control system at a specific time as the load dictates.

Pressure switches can be used for different purposes. Pressure switches are often used as safety devices to cut the equipment off if the pressure is dangerously high or low. In some cases in refrigeration systems, pressure switches are used as operating controls. Whatever the purpose of the pressure switch, it always reacts to a specific pressure in a certain way.

Transformers are used to break down the incoming power voltage to a voltage that can be easily used for control circuits.

Other control devices that we will discuss in this chapter include humidistats, oil safety switches, time-delay relays, time clocks, and solenoid valves.

In this chapter we will look at each of the control devices introduced in the preceding paragraphs and will describe how they operate in the system. In the next chapter we will see how troubleshooting on all control devices in a system can be done.

10.1 TRANSFORMERS

The transformer in a heating or cooling system provides the low-voltage power source for the control circuit. It transforms line voltage to the suitable low voltage that is needed for the control system. Most residences and small commercial installations incorporate a 24-volt control system. The transformer for a residential unit, shown in Figure 10.1, is used to transform the voltage from line voltage to 24 volts. Some commercial and industrial high-voltage equipment will use transformers like the one shown in Figure 10.2 to drop voltage from the supply to a 230-volt or a 110-volt control system.

FIGURE 10.1 Transformer used in residential and small commercial systems

FIGURE 10.2 Transformer used in large commercial and industrial units

Operation

Transformers are stationary inductive devices that transfer electric energy from one circuit to another by induction. The transformer has two windings, primary and secondary. An alternating voltage is applied to the primary winding of a transformer and induces a current in the secondary winding.

Many different types of transformers are used in the industry. A step-down transformer induces a secondary voltage at a lower rate than the primary. This type of transformer is used for the power supply of a low-voltage system. A step-up transformer induces a secondary voltage at a rate higher than the primary. This type of transformer is used to boost the voltage.

Sizing Transformers

Transformers are like many other electric components. They are not 100% efficient. In other words, there is a loss between the primary and secondary windings. This loss must be considered when sizing transformers for a certain job. Transformers are rated in voltamperes (VA). System equipment must be considered in transformer sizing along with the transformer rating.

The selection of a transformer vitally affects the performance and life of electric components in heating, cooling, and refrigeration equipment. A transformer too small for the control circuit will result in a lower-than-normal low voltage to the control circuit. This will result in improper operation of contactors or motor starters due to chattering or sticking contacts, burned holding coils, or the failure of contacts to close properly. All these conditions can cause system failure and possible damage to the compressor.

Even when transformers are sized correctly, care should be taken to avoid an excessive voltage drop in the low-voltage control circuit. When using a 24-volt control system with a remote thermostat, size the thermostat wire to carry sufficient current between the transformer and the thermostat.

The capacity of a transformer is described by its electric rating, primary voltage, frequency, secondary voltage, and the load rating in voltamperes. The maximum load of a transformer used in an air-conditioning control circuit is 100 VA. The maximum voltage is 30 V. Fuses can be used on the secondary side of the transformer for protection.

Transformers should be selected so they will operate all 24-volt loads without overheating. Transformers used only on furnaces will generally be rated 20 VA or less because of the light low-voltage loads. Air-conditioning systems usually require a transformer rated at 40 VA or larger, depending on the low-voltage loads. When replacing transformers in the field, use the following guidelines:

1. For replacement select a transformer the same size or larger than the one being replaced.
2. For new applications follow the manufacturer's recommendation.

Troubleshooting

Transformers can be easily diagnosed by two methods. An ohmmeter can be used to check the continuity of the windings of a transformer for an open, shorted, or good condition. However, it is difficult to diagnose a transformer with a spot burnout unless the second method of diagnosis is followed. A voltmeter can be used to check the secondary voltage of a transformer with the correct line voltage applied to the primary. When checking with a voltmeter, a load should be applied to the transformer. In some cases when a load is applied, the transformer secondary drops below an acceptable level, but this condition is rare.

When replacing a faulty transformer, the low-voltage control should be checked, because a short in the circuit will cause the transformer to burn out again. This is a common occurrence when a contactor coil or relay coil shorts out.

10.2 THERMOSTATS

The temperature in any structure, regardless of its age, location, or design, can be maintained at comfortable levels with a thermostat. Thermostats are designed and built in many different forms and sizes to meet the applications required in the industry. Thermostats play an important part in the total operation of most systems in the industry.

Applications

The basic function of a thermostat is to respond to a temperature change by opening or closing a set of electric contacts. Many different types of thermostats are used in the industry that perform a variety of switching actions. Figure 10.3 shows several common thermostats in use today.

Thermostats are used for many different purposes. An air-conditioning or heating thermostat would basically control the temperature of a given area for human comfort. Refrigeration thermostats are designed to maintain a specific temperature within a refrigerated space, such as in a domestic refrigerator, walk-in cooler, display case, and commercial freezer. There are many types of special application thermostats used in the industry, such as outdoor thermostats and safety thermostats. Whatever use a thermostat is put

FIGURE 10.3 Some common types of thermostats

to, it serves the same function: reacting to temperature with the opening and closing of a switch.

A heating thermostat closes on a decrease in temperature and opens on an increase in temperature. A cooling thermostat closes on an increase in temperature and opens on a decrease in temperature. This is a very important factor to consider when ordering or installing thermostats. A heating *and* cooling thermostat is available where heating and cooling are required. The heating and cooling thermostats are usually built so that the switch reacts either for heating or for cooling, with no intermediate position; in other words, a single-pole-double-throw switch. Some thermostats must isolate the heating and cooling contacts and therefore must use a separate set of contacts for heating and for cooling. Modern thermostats have a system switch that will determine whether the unit is heating or cooling.

Controlling Elements and Types of Thermostats

Two types of **thermostat controlling elements** are commonly used. The controlling element of a thermostat is the part that moves when a change in temperature is sensed. The bimetal thermostat shown in Figure 10.4 is com-

FIGURE 10.4 Internal parts of a thermostat

monly used to control the temperature of air in an air-conditioning or heating application. The remote bulb thermostat is commonly used to control the temperature of any medium, whether liquid or vapor, in many applications.

Remote Bulb Thermostat.   The remote bulb thermostat is shown in Figure 10.5(a). The power element, which is the bulb, and the diaphragm are interconnected with a section of small tubing. The bulb is filled with liquid and gas and then is sealed. The pressure exerted by the diaphragm on the mechanical linkage will open and close a set of contacts. A diaphragm with its remote bulb is shown in Figure 10.5(b). As the bulb temperature changes, so will the pressure exerted on the diaphragm. If the temperature of the bulb increases, so will the pressure. If the temperature of the bulb decreases, so will the pressure. The increase or decrease in pressure causes the contacts to open or close, depending on the design of the thermostat.

Bimetal Thermostat.   The heart of most other types of thermostats is a

| (a) Thermostat | (b) Power element and diaphragm |

FIGURE 10.5 Remote bulb thermostat

bimetal element. The element gets its name from the fact that it uses a bimetal to cause the movement that opens and closes a set of contacts. A bimetal is a combination of two pieces of metal, as shown in Figure 10.6(a) that are a given length at a certain temperature. The two metals are welded together. Each metal has a different coefficient of expansion. If the temperature of these two metals is increased, one will become longer than the other because of the different expansion qualities. This causes the bimetal to arch, as shown in Figure 10.6(b). If the bimetal is anchored at one end, leaving the other end to move freely, it will move up and down according to the temperature surrounding it. In a low-voltage thermostat, the bimetal works better and gives better control if it is large.

The first type of bimetal thermostat produced is shown in Figure 10.7(a). It was unsatisfactory because of the unstable pressure from the bimetal that held the contacts together. It would react (close or open) to a relatively small change in the temperature around the room thermostat.

A thermostat must have a means of making a good connection with the contacts. This is accomplished by a **snap action** of the thermostat bimetal to the fixed contacts. The early type of thermostat was not snap-acting in the switching movement and thus caused problems because of its reaction to minor temperature change. However, if a permanent magnet is placed near the bimetal arm, as shown in Figure 10.7(b), it will cause a snap action when the bimetal expands enough to move the contacts close together.

Two methods in common use enable thermostats to be snap-acting: a permanent magnet and a mercury bulb. A permanent magnet mounted near the fixed contacts will cause the action of the bimetal to be snap-acting. The bimetal will make good contact once it is in the magnetic field of the permanent magnet. Figure 10.8 shows this type of bimetal thermostat.

The mercury bulb thermostat also provides snap action because of the globule of mercury moving between two probes sealed inside a glass tube,

(a) **Two metals with different coefficients of expansion**

(b) **Metal A expands more than B and causes an arch**

FIGURE 10.6 A bimetal element for a thermostat

(a) **Basic bimetal with no means of snap action to produce a stable set of contacts**

(b) **Bimetal with permanent magnet to produce snap action of the contacts**

FIGURE 10.7 Basic bimetal elements

Contacts

Bimetal

Temperature
Dial

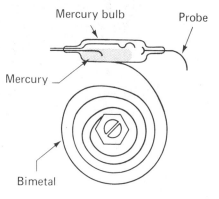

Mercury bulb

Probe

Mercury

Bimetal

FIGURE 10.8 Bimetal thermostat with snap-action
contacts

FIGURE 10.9 Bimetal with a mercury bulb used as
the contacts

as shown in Figure 10.9. Figure 10.10 shows a bimetal thermostat with a
mercury bulb that is used in the industry.

Line Voltage Thermostat. The line voltage thermostat is designed to operate on line voltage, for example, 110 volts or 230 volts. The line voltage
thermostat, shown in Figure 10.11, is used on many packaged air-conditioning
units and refrigeration equipment of commercial design. This type of thermostat is used to open or close the voltage supply to a load in the system.
Figure 10.12 shows the wiring diagram of a common 230-volt window air
conditioner with a line voltage thermostat.

The line voltage thermostat lacks many good qualities that are obtained
with low-voltage thermostats. For example, depending on the application,
the line voltage thermostat functions only to open or close a set of contacts.
Line voltage thermostats are commonly used in the industry.

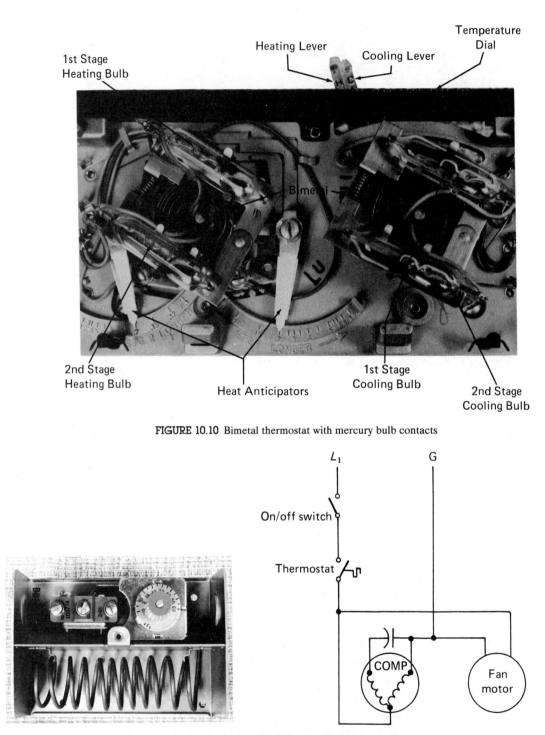

FIGURE 10.10 Bimetal thermostat with mercury bulb contacts

FIGURE 10.11 Line voltage thermostat

FIGURE 10.12 Wiring diagram of window unit with line voltage thermostat

FIGURE 10.13 Wiring diagram of a residential heating and cooling unit with a low-voltage thermostat

Low-Voltage Thermostat. The low-voltage thermostat is used on control systems with a 24-volt supply. The low-voltage thermostat is used on all residential heating and air-conditioning systems and many commercial and industrial systems. The low-voltage thermostat can be used for heating operation, cooling operation, automatic operation of fans, manual operation of fans, and automatic changeover from heating to cooling. The difference between a line voltage and a low-voltage thermostat is in the size of the bimetal element. With the larger line voltage contacts, more pressure is needed to close the contacts and therefore larger bimetal elements are required. Figure 10.13 shows a wiring schematic of a residential heating and cooling system with a low-voltage thermostat. The low-voltage thermostat is more accurate, less expensive, and requires smaller wiring than its counterpart, the line voltage thermostat.

Anticipators

The success of any thermostat depends on the system being correctly sized, the air flow being balanced, and the thermostat being correctly placed. It is almost impossible for thermostats to maintain the exact temperature of any given area without spending an unreasonable amount for controls. Thus, **anticipators** are used to give a more evenly controlled temperature range. There are two types of anticipators: heating and cooling. We will discuss each type in turn.

Heating Anticipator. A thermostat that has no means of heat anticipation will allow a wide swing from the desired temperature, especially on forced-warm-air systems. If the thermostat is set on 75°F, the furnace will come on for heating. But there is a delay before any warm air is carried to the conditioned space because a warm-air heating system must heat the furnace first. This delay will allow the air temperature to drop to 74°F or less before the blower begins operation and heat is carried to the conditioned space.

The temperature difference between the closing of the thermostat and the time when warm air begins to reach the thermostat is called **system lag**. As long as the furnace remains on, the temperature will continue to rise. If the thermostat differential is 2°F, the thermostat will open at 77°F and stop the burner. However, the furnace is still warm and the blower must be operated until the furnace cools, which will carry the temperature to 78°F or higher.

The temperature difference between the closing of the thermostat and the time when the warm air is no longer being delivered to the room is called **overshoot**. The overshoot and system lag can produce an additional temperature swing of as much as 5°F. Figure 10.14 shows the effect of overshoot and system lag.

The wide difference in temperature can be controlled by using a heating anticipator in the thermostat. The heating anticipator is nothing more than a small source of heat close to the bimetal element. This allows the bimetal to be a little warmer than the surrounding air. The heating anticipator is placed in series with the contacts of the thermostat. When the contacts close and energize the burner, the current must also flow through the heating anticipator, which causes it to heat the bimetal to a certain temperature.

A heating anticipator anticipates the point at which the thermostat should open and provides a narrow thermostat differential. Suppose the thermostat is set at 75°F and the furnace burner is off, with the temperature dropping slowly. When the temperature falls below 75°F, the burner starts and at the same time the heat anticipator begins to heat the bimetal. The air tempera-ture will continue to drop until the blower delivers warm air to the condi-

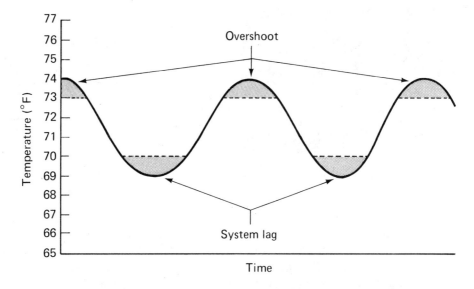

FIGURE 10.14 Effect of overshoot and system lag on room temperature

tioned space. The room temperature begins to rise, but the bimetal temperature is slightly higher because of the heat from the heating anticipator. Thus, the thermostat anticipates the overshoot and cuts off the burner. Due to the shorter length of time that the burner operates, the furnace will cool quicker, stopping the fan and preventing the overshoot. A heating anticipator does not eliminate system lag and overshoot. But these factors become negligible with a heating anticipator, producing a temperature swing of only 2°F to 3°F. The thermostat circuit previously shown in Figure 10.13 shows a heating anticipator connected in series with the contacts of the thermostat.

Two types of heating anticipators are used: fixed and adjustable. The fixed heating anticipator is nonadjustable and is not versatile. The adjustable heating anticipator, shown in Figure 10.15, can be matched to almost any control system. The adjustable heating anticipator should be set on the current draw of the primary control or the gas valve. For example, if a gas valve had a current draw of .2 amps, then the heating anticipator should be set on .2 amps. Matching the adjustable heating anticipator to the current rating of the burner control assures the best possible heat anticipation.

Cooling Anticipator. The cooling anticipator operates somewhat differently from the heating anticipator. This type of anticipator is also known as an off-cycle anticipator. The cooling anticipator is shown connected to a thermostat in Figure 10.16. The cooling anticipator adds heat to the bimetal on the "off" cycle of the equipment because of its parallel connection in the

FIGURE 10.15 Adjustable heating anticipator

FIGURE 10.16 Schematic diagram of cooling anticipator connected to a thermostat

circuit. When the thermostat contacts close, the current takes the flow of least resistance, which is through the contacts rather than the cooling anticipator. On the "off" cycle the current passes through the anticipator and the contactor coil. The anticipator drops the voltage to a point that will not energize the contactor, because of the series connection. The current flow through the anticipator heats the bimetal and causes the thermostat to anticipate the temperature rise. The cooling anticipator is not as important as the heating anticipator because cooling equipment operation is almost instant, while there is a delay in the heat delivery of a forced-warm-air system.

The thermostat is merely a switching device that routes the voltage to the correct control for the operation prescribed by the thermostat setting. Figure 10.17 shows the schematic of a low-voltage thermostat with its routing. With the system selector in the cooling position, voltage flows through the system switch to the cooling thermostat, which operates the cooling equipment and the fan motor to maintain the temperature setting. If the system switch is in the heating position, the current flows through the switch to the heating thermostat, which operates the heating equipment. Many thermostats incorporate a fan switch that allows the fan to be operated manually or with the cooling equipment, since the fan on heating operates from a fan switch.

Thermostat Installation

Installation of a thermostat is simple because all thermostats are marked with identifiable letters, although the letters used are not consistent from manufacturer to manufacturer. Figure 10.18 on page 226 shows a simple heating and cooling thermostat subbase. The terminals are shown with their letter designations (the thermostat attaches to the subbase). The letter designations to system functions are given in the chart in Figure 10.18. The letter designation should be followed for proper installation.

When checking thermostat operation, check the wiring connections and make sure that the selector switch and the temperature setting are properly selected.

When installing a thermostat, choose a suitable location, which helps maintain the correct temperature in the desired area. Thermostats installed in

⚠1 With no O terminal load, thermostat current during heating cycle varies depending on whether fan switch is in the On or Auto position. Heater should be set for combined current level of heat relay and fan relay coils. With O terminal load, set thermostat heat anticipator to its maximum setting as cooling anticipator in series with O terminal load provides heat anticipation in the heating cycle. (Limit the thermostat heating load current to 0.8 amps to assure good performance.)

⚠2 Power supply provides overload protection and disconnect means as required.

FIGURE 10.17 Schematic diagram of the routing of a low-voltage thermostat *(Courtesy of Honeywell)*

Letter Designation

Voltage to Transformer	R–V
Heating	W–H
Cooling	Y–C
Fan	G–F
Heating Damper (seldom used)	B
Cooling Damper (seldom used)	O

FIGURE 10.18 Thermostat subbase with terminals and letter designations

applications that require control of the temperature of some enclosed area, such as a refrigerator, walk-in cooler, and walk-in freezer, or some medium such as water in a chilled water system, should be installed where the correct temperature can be sensed by the thermostat without interference from some other source of heat or cold. Thermostat installations in commercial refrigeration applications should be where the thermostat can sense the temperature that reflects the average temperature of the box or case. When installing thermostats in applications where some medium such as water is being controlled, mount or insert the thermostat bulb or controlling element in the system so that the correct temperature is being sensed by the thermostat. When replacing a thermostat on a system, make certain that the thermostat is the correct type and that it is installed in the same way as the one being replaced. The thermostat should not be installed where there is a possibility of interference by some installation.

The thermostat in an air-conditioning application should be located in the conditioned space on a solid inside wall where there is normal air circulation and where it will not be subject to any artificial heat or cold from lamps, televisions, appliances, or hot water pipes. The thermostat should be installed about 5 feet above the floor.

Troubleshooting

Troubleshooting thermostats will be a common job for installation and service mechanics in the industry. Most thermostat manufacturers make available detailed installation and service procedures for their thermostats. When troubleshooting thermostats, the mechanic will be involved in three basic areas: calibration, diagnosis, and maintenance.

A common complaint is that the thermostat does not turn the equipment on and off at the proper temperature or that the temperature does not agree with the thermometer on the thermostat. Calibration is a means of resetting the thermostat so that the temperature more accurately reflects the temperature of the structure. When calibrating thermostats, follow the manufacturer's recommendations.

Common complaints about thermostats are excessive temperature swing, the system short cycles, the system on cycle is too long, or the thermostat works incorrectly in regard to heating and cooling or no action. In many cases that involve temperature swing and cycling time, correctly setting the heat anticipator, if the thermostat has one, will correct the problem. Most line voltage thermostats do not have heat anticipators. Other problems often encountered by service mechanics are that the thermostat is not doing what it is designed to do, or that the thermostat is defective and must be replaced.

Maintenance of thermostats covers checking the terminals for tightness, cleaning the contacts if possible, checking the calibration, and cleaning the sensing element.

10.3 STAGING THERMOSTATS

A **staging thermostat** is designed to operate equipment at different times with respect to the equipment's needs (Figure 10.10). The staging thermostat has more than one contact and opens and closes at different times with regard to the condition of the area being controlled.

The heating, cooling, and refrigeration industry has advanced in most phases, but in recent years there has been a demand for better control and more efficient operation with the modern systems. Staging thermostats have been designed to meet this need. The staging thermostat is becoming increasingly popular in the industry because of its versatility in system control.

Staging System

Many heating and cooling systems are operated in stages because the load in some structures fluctuates a great deal. A heating or cooling system that has been designed to operate on two different capacity levels is a **staged system**. A staged system is designed to operate at half of its capacity or more until the operating section of the equipment can no longer handle the heating or cooling needs of the structure. Then additional stages are called on as they are needed.

Staging systems offer many advantages because of their more efficient operation. For example, on a mild day the cooling load of a building is low and the full capacity of the equipment is not needed. Hence, the equipment will operate at half of its capacity. If the day is extremely hot and the full capacity of the system is needed, then it can be utilized by use of a staging thermostat. Staging can be used to many advantages whether it is in heating or cooling.

Operation and Types

A staging thermostat is designed to be used on a system that has two stages of heating, cooling, or both. The staging thermostat operates on the differential in temperature between the stages. For example, a two-stage-cooling thermostat could close one set of contacts at 75°F and the other set at 76.5°F to 78°F. A two-stage-heating thermostat could close one set of contacts at 75°F and the other set at 73.5°F to 72°F. This allows systems to operate on partial capacity until the need arises for full capacity.

Staging thermostats can be obtained in a variety of stage configurations. Common staging thermostats used in the industry are the one-stage-heating thermostat with two-stage-cooling, the two-stage-heating thermostat with one-stage-cooling, and the two-stage-cooling thermostat with with two-stage-heating. Figure 10.19 shows a two-stage-heating and two-stage-cooling thermostat. Note that four mercury tubes are used, with two mounted on the cooling control level and two mounted on the heating control level. Figure 10.20 shows the schematic diagram for the two-stage-heating and two-stage-cooling thermostat.

The letter designations are the same for staging thermostats as they are for regular thermostats except that heating or cooling stages would be lettered with a 1 or 2 after the main letter. The number 1 represents the first stage; the number 2 represents the second stage. When installing staging thermostats, pay careful attention to the letter designations and control hookup to avoid an improper operating system.

Heat Pump

A **heat pump** is a refrigeration system that heats or cools by reversing the refrigerant cycle for the heating operation and then operates conventionally for cooling. Most heat pumps use a staging thermostat to operate a set of supplementary heaters. The first-stage thermostat operates the compressor. The second stage operates the supplementary heat when the first stage cannot handle the load. Figure 10.21 is a wiring diagram of a heat pump showing the two stages of heating and what they control.

FIGURE 10.19 Low-voltage thermostat for two-stage cooling and two-stage heating

FIGURE 10.20 Schematic diagram for a two-stage-cooling and two-stage-heating thermostat *(Courtesy of White-Rodgers Division of Emerson Electric Company)*

FIGURE 10.21 Schematic diagram of a heat pump using a two-stage-heating and one-stage-cooling thermostat to control the unit

10.4 CLOCK THERMOSTATS

A **clock thermostat** is used to control the temperature of a structure, allowing the consumer to lower or raise the temperature control point for at least one period in every 24 hours. Figure 10.22 shows a common clock thermostat. With energy conservation playing a larger part in air-conditioning and heating systems, the clock thermostat is becoming increasingly popular. Clock

FIGURE 10.22 Clock thermostat

thermostats give the consumer the option of lowering the temperature for a specified period of time.

Applications

Clock thermostats are being widely used to allow a homeowner to set the temperature down or up for at least one period in 24 hours. The clock thermostat would be used in a structure to allow for a setback temperature. This setback operation is usually done at a period of time that would least affect the homeowner. In most cases, the temperature would be set back at night when all occupants are sleeping and would be automatically reset to the desired level before the occupants needed the increased temperature.

The most practical application of clock thermostats is generally for heating systems, but they are available for cooling systems as well. There are many different types of clock thermostats being produced today, and there is almost no limit to the functions they can perform.

Several different types of clock thermostats are used today. There are single-setback and multisetback clock thermostats available, with a wide variety in price. Some clock thermostats use a 24-volt power supply that must be installed, while others use batteries mounted in the thermostat. A solid-state clock thermostat is available with completely programmable setback temperatures.

There are three basic types of clock thermostats. The simplest type of clock thermostat is the single-setback, which allows an adjustable program to automatically lower or raise the temperature control point for one period in every 24 hours. The pickup and setback times need to be set only once for continuous operation of the system. This type of clock thermostat is supplied by an external 24-volt power supply. Figure 10.23 shows a single-setback clock thermostat with the cover removed. Figure 10.24 shows the correct procedure for setting this type of clock thermostat.

The multisetback programmable clock thermostat allows for more than one automatic lowering or raising of the temperature control point several times in a 24-hour period. Figure 10.25 shows a multisetback program clock thermostat. This type of thermostat is programmed by properly setting levers

FIGURE 10.23 Single-setback clock thermostat with the cover removed

FIGURE 10.24 Setting procedure for a single-setback clock thermostat

FIGURE 10.25 Multisetback programmable clock thermostat

on the thermostat. A manual program override will allow the bypass of a setback temperature. This type of clock thermostat is available with its own battery as a power supply. Models utilizing a separate 24-volt power supply are also available.

The most sophisticated type of clock thermostat is a solid-state model, shown in Figure 10.26. The solid-state clock thermostat allows for a multiple setback with the option of a 5- or 7-day program, which allows the homeowner to omit weekend setbacks while the structure is occupied. This type of thermostat is equipped with a digital clock. Figure 10.27 shows the thermostat with the cover removed. Programming this thermostat is simple and easy.

FIGURE 10.26 Solid-state clock thermostat

FIGURE 10.27 Solid-state clock thermostat with the cover removed

Installation

Installation of a clock thermostat with regard to the temperature function is the same as that of any low-voltage thermostat. The letter designations are also the same as those for other low-voltage thermostats. Figure 10.28 shows the diagram of a clock thermostat used for a heating application; the thermostat is shown in the shaded area. Figure 10.29 shows a heating-cooling clock thermostat with subbase; the shaded area indicates the thermostat. For a clock thermostat that requires an external 24-volt power supply, connections must be made at a separate power supply, as shown in Figure 10.30, or to the 24-volt power supply of the system. Clock thermostats using batteries require no additional power wires.

10.5 PRESSURE SWITCHES

A pressure switch is a device that opens or closes a set of contacts when a certain pressure is applied to the diaphragm of the switch. A high-pressure

FIGURE 10.28 Wiring diagram of heating clock thermostat *(Courtesy of Honeywell)*

FIGURE 10.29 Wiring diagram of a heating-cooling clock thermostat, where shaded area indicates the thermostat and white area inside the block represents the subbase *(Courtesy of Honeywell)*

switch is connected to the discharge side of the system to sense discharge pressure. A low-pressure switch is connected to the suction side of the system to sense suction pressure. Pressure switches can be used as safety devices, as main operating controls, or to operate other parts of the system. Figure 10.31 shows a common pressure switch used in the industry.

Two types of pressure switches are used in the industry today. A nonadjustable pressure switch is used by many manufacturers to prevent the pressure setting from changing. An adjustable pressure switch can be adjusted to meet any specific need that might arise. Adjustable pressure switches are

FIGURE 10.30 A 24-volt power supply for a clock
thermostat

(a) With enclosure　　　　　　　　**(b) Without enclosure**

FIGURE 10.31 Pressure switch

usually obtained for field replacement of a pressure switch. Pressure switches can also be classified as high pressure or low pressure. These switches open or close on a rise or fall of pressure. Pressure switches are manufactured with a case enclosing the switch, as shown in Figure 10.31(a) and for mounting in a control panel without the case, as shown in Figure 10.31(b). A dual-pressure switch contains both a high-pressure and a low-pressure switch, as shown in Figure 10.32. Pressure switches can be obtained for most any purpose that would arise in the industry.

High-Pressure Switch

A high-pressure switch is usually used as a safety device to protect the compressor and system from excessively high discharge pressure. A high-pressure switch used as a safety control must open on a rise in pressure to shut the equipment down.

The high-pressure setting of the switch must correspond to the type of refrigerant in the system. The setting of a high-pressure switch used with Freon 12 would be different from the setting of a high-pressure switch used with Freon 22. Figure 10.33 lists the common setting points of a high-pressure switch for the common refrigerants used in the industry today. High-pressure

FIGURE 10.32 Dual-pressure switch (combination of high and low pressure)

| | | High-Pressure Setting | |
Refrigerant	Condenser Type	Cut-out	Cut-in
12	Air Cooled	225	145
	Water Cooled	170	90
500	Air Cooled	280	200
	Water Cooled	210	130
22	Air Cooled	380	300
	Water Cooled	280	200

FIGURE 10.33 Table of approximate setting points of high-pressure switch used as a safety control

switches may also close on a rise in pressure to operate a device that would control the discharge pressure.

Low-Pressure Switches

Low-pressure switches are used as safety devices, as operating controls, and as devices to operate any component by the suction pressure of the system. All low-pressure switches are connected to the suction side of a refrigeration system. Low-suction pressure can cause damage to the compressor. Therefore, low-pressure switches are used many times as a safety device to prevent damage to the system when the suction pressure drops below a predetermined point.

Figure 10.34 lists the common setting points of a low-pressure switch for different types of Freon.

If the low-pressure switch opens, it should break the control circuit that operates the compressor. Low-pressure switches can also be used as an operating control to operate the system by a pressure setting that corresponds to a temperature setting. Therefore, by properly setting a low-pressure switch, the temperature can also be controlled. Figure 10.35 lists the proper pressure control settings for specific applications.

Notation and Terms

Many terms should be understood to successfully maintain and set pressure switches. The **differential** of a pressure switch is the difference between the cut-in and cut-out pressure of the switch. The **cut-in** is the pressure of the system when the pressure switch closes. The **cut-out** is the pressure of the system when the pressure switch opens. The **range** of a control is the operating range of the system—for example, the overall pressure—over which the switch can operate. These terms are used frequently when setting pressure to obtain the correct operation of the system. Many pressure controls

	Low-Pressure Setting	
Refrigerant	Cut out	Cut in
12	15	35
500	22	46
22	38	68

FIGURE 10.34 Table of approximate setting points of a low-pressure switch used as a safety control

| | REFRIGERANT | | | | | |
| | 12 | | 22 | | 502 | |
Application	In	Out	In	Out	In	Out
Walk-in Cooler	32	15	60	35	72	40
Dairy Case–Open	35	12	64	24	75	30
Reach-in Display Case–Open	36	14	66	33	78	40
Meat Display Case–Closed	35	15	65	34	78	45
Meat Display Case–Open	30	12	55	30	65	35
Vegetable Display–Open	40	15	80	36	90	45
Beverage Cooler–Wet Type	30	20	58	38	68	58
Beverage Cooler–Dry Type	35	15	65	35	76	43
Florist Box	45	28	80	55	85	63
Frozen Food–Open	6	6	16	4	25	10
Frozen Food–Closed	10	2	22	12	30	15
Walk-in Freezer (–10°F)	12	1	32	10	25	15

Caution: The figures are approximate and should be used only as a guide. Special applications should not be considered.

FIGURE 10.35 Table of approximate pressure control settings of a low-pressure switch used as an operating control

can be set by carefully reading the pressure dial of a pressure switch. Care should be taken when setting or replacing switches.

Troubleshooting

There are many problems commonly encountered when working with pressure switches. The contacts of pressure switches frequently cause problems because of pitting, wear, mechanical linkage faults, and sticking. The contacts should be checked to see if they are closed or open in relation to the setting of the control.

If the contacts show continuity, the resistance should also be checked. If the resistance is above allowable limits, or if the contacts prove to be faulty, the pressure switch should be replaced. If any resistance is read through a set of contacts, voltage is being lost. A reading of 2 or 3 ohms of resistance in a set of switches indicates that the switch should be replaced or at least that the voltage should be checked across the contacts.

The pressure connections of a pressure switch can leak and cause a faulty control or no control at all. To check for a leak use an acceptable method for locating Freon leaks. Never replace a pressure switch before its correct position and purpose have been determined.

10.6 MISCELLANEOUS ELECTRIC COMPONENTS

Humidistats

In certain air-conditioning applications, it is essential to control the humidity. **Humidistats** are used to control the humidity of a structure. The humidistat uses a moisture-sensitive element to control a mechanical linkage that will open or close an electric switch according to the humidity. Figure 10.36 shows a humidistat enclosed and with the cover removed.

In certain fiber and cotton industries, the humidity must be accurately controlled by a humidistat, which controls the operation of air washers to prevent damage to the finished product. Humidity control is also advisable in the winter in many structures to produce a more comfortable area by adding more moisture to the air.

Oil Safety Switch

Oil safety controls are essential on commercial and industrial equipment to provide adequate protection for the compressor in case the oil pressure drops. Large compressors use a pressure type of lubrication system that must be maintained at a given pressure to ensure proper lubrication of the compressor.

To correctly read the oil pressure of a compressor, you must subtract the suction pressure from the oil pressure to get the net oil pressure, because

(a) Enclosed   **(b) Without cover**

FIGURE 10.36 Humidistat

the suction pressure exerts pressure on the oil in the crankcase. Therefore, the pressure connections of an oil safety switch should be connected to the oil pressure port on the compressor and to the suction or crankcase pressure. These connections of pressure to the oil safety switch will transfer the net oil pressure to the control by some type of mechanical linkage to open and close a set of contacts, depending on the oil pressure setting. Figure 10.37 shows an oil safety switch and its schematic diagram.

The oil safety switch is designed to allow a certain time delay so that the oil pressure will build up in the compressor after the start-up of the compressor. When the compressor is energized, the time-delay switch in the oil safety control is also energized. If the oil pressure does not reach a certain level within the time-delay period, the control circuit will be deenergized. However, if the oil pressure reaches the desired level, the time-delay switch is removed from the circuit and the compressor continues to operate. The time-delay switch is nothing more than a heater that opens a bimetal switch after the specified period of time delay.

Oil safety switches are rated for pilot duty and in most cases must be manually reset. Oil safety controls are essential on large expensive compressors to prevent undue damage because of inadequate lubrication.

Time-Delay Relay

Time-delay relays are used in the industry to delay the starting of some load for a designated period of time. Time-delay relays usually use some type of heating element that closes a bimetal element. The time-delay function is brought about by the period of time that it takes the heating element to open or close the bimetal. Two types of time-delay relays are shown in

(a) Switch   (b) Schematic

FIGURE 10.37 Oil safety switch

FIGURE 10.38 Time-delay relays

FIGURE 10.39 Clips attached to the dial of a one-day time clock

Figure 10.38. The relay contacts are usually rated as pilot duty. The coil or heater of the relay can be rated at 24 volts, 100 volts, or 230 volts.

Time-delay relays can be used to prevent two heavy loads of a system or two units connected in parallel from starting at the same time by putting the contacts of the time-delay relay in series with the load or controlling element of the unit. Some commercial and industrial units use a part winding motor, which uses a time-delay relay to energize the windings at different times. The delay period of a time-delay relay can vary from 15 seconds to 30 seconds or longer.

Time Clock

Time clocks are devices that open and close a set of contacts at a selected time by a mechanical linkage between the contacts and a clock. Time clocks can operate on a one-day basis or a seven-day basis. The one-day time clock, shown in Figure 10.39, operates on a 24-hour period of time without regard to the day of the week. A seven-day time clock operates on an hourly or a daily basis during a week.

Time clocks are set by attaching clips to the clock wheel or dial, as shown in Figure 10.39. There are two clips that attach to the wheel or dial. On movement of the dial, one clip closes the contacts at the point in time where it is attached to the dial. The other clip opens the contacts at the point in time where it is set on the dial.

Time clocks are used to operate equipment or sections of equipment on a time basis. One common use of a time clock is to stop and start the defrosting cycle of a refrigeration system. Time clocks are also used to cut systems off

FIGURE 10.40 A common solenoid valve

when a structure is to be unoccupied and to cut systems in before building occupants arrive. Buildings such as churches, offices, or manufacturing plants that close and open at specific times often use time clocks. Time clocks can be used effectively to save the operating cost of equipment when it is not needed.

Solenoid Valves

Solenoid valves are valves that open and close due to a magnetic or solenoid coil being energized, which pulls a steel core into the magnetic field of the solenoid. A common solenoid is shown in Figure 10.40. Solenoid valves stop or start the flow of a fluid such as water, air, or refrigerant. Solenoids may be normally open or normally closed and care should be taken to identify the normal position of the valve.

In this section only a few miscellaneous electric controls have been covered, but many others are used in the industry for special purposes. Most controls not covered can be understood by studying the device and reading the manufacturer's instructions on the device. Service personnel should understand all basic electric components and their operation to correctly diagnose problems in components. Some devices are difficult and complex but in most cases manufacturers will assist mechanics in becoming familiar with special devices by their literature and through service schools.

SUMMARY

In this chapter we discussed several different types of control devices, among them thermostats, pressure switches, and transformers.

Thermostats are designed to open and close a set of electric contacts. They are used to control temperature in a structure or the temperature of a medium such as a liquid or gas. There are two types of thermostats used in the industry today. The remote bulb thermostat is used for control of any medium and is usually used in commercial and industrial applications. The bimetal thermostat is used in all residences and in many cases when air temperature is controlled in commercial and industrial applications. Thermostats can also be of the line voltage or low-voltage type. The line voltage thermostat is used to open or close the voltage supply to a load in a system. The low-voltage thermostat is often used as a switching device to control low-voltage loads.

The heating thermostat that does not have a heating anticipator will allow the temperature of the controlled area to have a large swing in differential. Heating anticipators, when incorporated with a heating thermostat, will have a small temperature differential. Heating anticipators are connected in series with the contacts of the thermostat and set to match the burner control. Cooling anticipators are connected in parallel and operate when the equipment is off. They anticipate the need of the equipment to operate sooner because of system lag. Thermostats play an important part in the industry because of their function of temperature control.

Staging thermostats are used to provide a means of controlling two stages of heating or cooling with a set differential. The staging of equipment gives better comfort and more efficient operation of equipment. Staging is used exclusively in commercial and industrial applications.

A clock thermostat is used to control the temperature of a structure. It allows the consumer to lower or raise the temperature control point for at least one period in 24 hours. The clock thermostat gives the homeowner the flexibility of setting back the temperature in his or her home when the family is sleeping, for example, or when the family is away from home.

Pressure switches are widely used devices in the industry. They are often used to control a load in a system. Pressure switches are designed to open or close on a rise or fall in pressure, depending on their application.

Low-pressure switches are often used in the industry to protect refrigeration components from low-suction pressure. Low-pressure switches are used to operate systems or components and to act as safety devices to protect the refrigeration components of the system.

High-pressure switches are used as safety devices to protect the refrigeration system from excess pressure or to operate a component for head pressure control. High head pressure can damage a compressor, metering device, and high-side system piping. High-pressure switches can also be used to control loads that must be started from the discharge pressure. Careful attention should be paid to the type of system and refrigerant when checking the pressure setting of a control.

Transformers decrease or increase the applied voltage to the desired voltage by use of induction. Transformer ratings include primary voltage and frequency, secondary voltage, and voltamperes. Capacity is rated in voltamperes, that is, the voltage times the current of the control circuit loads. The more low-voltage loads in a control circuit, the larger the transformer that must be used.

QUESTIONS

1. What is the purpose of a thermostat?

2. The two types of thermostats are _____ and _____.

3. What is the difference in the operation of a heating thermostat and a cooling thermostat?

4. The controlling element of a thermostat is the part _____.

5. Explain the operation of the remote bulb thermostat.

6. True or false. A bimetal is a combination of two pieces of metal welded together.

7. Why should the contacts of a thermostat be snap-acting?

8. Name the two methods in common use that enable thermostats to be snap-acting.

9. True or false. The line voltage thermostat reduces the line voltage being supplied to the load.

10. The low-voltage thermostat is used on _____.

11. The difference between a line voltage and a low-voltage thermostat is in _____.

12. What are the letter designations R, Y, Y1, Y2, W1, W2, and G used for in thermostats?

13. What is the purpose of a heating anticipator?

14. What is the proper procedure for checking thermostats?

15. Describe what is meant by system lag and overshoot.

16. What is the purpose of staging heating and cooling equipment?

17. Why are two-stage-heating thermostats used on heat pumps?

18. What is the purpose of a low-pressure switch?

19. What is the purpose of a high-pressure switch?

20. How can a low-pressure switch be used as an operating control on a refrigeration system?

21. Define the following terms: cut-in, cut-out, range, and differential.

22. What is the purpose of a transformer and how does it work?

23. True or false. A step-down transformer is used to boost voltage.

24. What is the correct procedure for checking transformers?

25. The capacity of a transformer is rated in _____.

26. What precautions should be taken when replacing transformers?

27. What is the purpose of a humidistat?

28. The oil safety control operates on what two pressures in the refrigeration systems?

29. Time-delay relays are used in heating, cooling, and refrigeration systems to _____ .

30. Give several reasons for using a time clock in the heating and cooling industry.

31. Why is it important for service personnel to understand the operation of any electric component in the system?

32. What is the purpose of a clock thermostat?

33. What is a multisetback clock thermostat?

34. How is the clock portion of a clock thermostat powered?

35. Explain the difference between a single-setback and a multisetback clock thermostat.

36. What routine maintenance should be performed on a thermostat controlling residential heating and cooling?

37. What is calibration with respect to a thermostat?

LAB SECTION REFERENCE

For experiments and activities dealing with material covered in this chapter, refer to Lab Number 8, Transformers; Lab Number 10, Thermostats; and Lab Number 11, Pressure Switches.

11

Troubleshooting Electric Control Devices

INTRODUCTION

Most troubleshooting in a system involves a specific problem that the consumer is encountering. Most problems in a system stem from one source—an electric component that is not functioning properly or is faulty. It is the responsibility of the service mechanic to locate the component that is not functioning correctly and to replace or repair it. Sometimes it is difficult to locate the exact trouble in the entire system, but the task should be relatively simple once the problem has been narrowed down to a single or several components. In this chapter we will discuss troubleshooting for most of the basic electric control components.

The first step in troubleshooting any component is to understand its operation and function. If the operation of an electric component is not understood, it is impossible to effectively check the component. Electric meters will usually be needed in the diagnosis of the component. Thus, it is essential for service mechanics to understand the use of electric meters. Service mechanics and other personnel must also understand the proper procedures for checking electric components and be able to correctly diagnose the condition of the component. In the following sections, we will discuss some guidelines to use in checking electric components and in diagnosing problems of components in modern heating, air-conditioning, and refrigeration systems.

11.1 ELECTRIC MOTORS

Electric motors are the most important loads in any heating, cooling, or refrigeration system. They are used almost exclusively to cause the rotating motion of fans, compressors, pumps, and dampers. Many different types of electric motors are used in the industry. However, the type of motor used will have no effect on diagnosing the condition of the motor but will have a great effect on the selection of a replacement motor if needed.

Open type electric motors will fail in three different areas: windings, centrifugal switch, and bearings. The windings can be checked for opens, shorts, and grounds. One important element in the diagnosis of an electric motor winding is to know the type of motor because of the winding layout. The condition of the centrifugal switch can best be determined by visual inspection after disassembling the motor; the operating characteristics can also tell much about the condition and operation of the centrifugal switch. The bearings can usually be checked by turning the motor by hand to determine if hard or rough spots are in the rotating movement of the motor.

The only part of a sealed motor, such as a hermetic compressor, that can be checked is the windings because there are no internal parts other than the bearings. The winding can be checked as any other motor for opens, shorts, and grounds. Diagnosis of bearing failure in a hermetic motor is often difficult because no visual or feeling of the rotation can be accomplished. Bearing failure on hermetic motors must be diagnosed through amperage readings of the motors along with running characteristics. However, several different types of starting apparatus need to be checked. It is imperative that service mechanics be able to diagnose the condition of the many types of starting relays and start assist devices used in the industry. These starting components were discussed in Chapter 8.

Motor replacement procedure is important to the mechanic in the field. Many times the exact replacement motor is not available, and the mechanic must adapt a different motor to meet the specifications. Replacement motors can sometimes be identified by a manufacturer's part number or by the model number of the unit, but in many cases identification must be made by the motor and its nameplate. The mechanic must have enough information to identify a motor that will replace the defective one. Many unknown elements such as type of motor, rotation, number of motor speeds, and horsepower can be identified by the mechanic when removing the faulty motor from the unit. The mechanic must be sure the replacement motor selected will operate properly for the application that it is being used.

Troubleshooting motors was covered in detail in Chapter 7.

11.2 CONTACTORS AND RELAYS

Contactors and relays are used on most heating, cooling, and refrigeration equipment for the operation of loads in the system. Contactors and relays are similar in their operation because both contain sets of contacts and a coil used to open or close the contacts. The contactor is larger and capable of carrying more amperage than the relay.

The same procedure can be used to check both contactors and relays. Three areas of problems are encountered with contactors and relays: contacts, coil, and mechanical linkage. Any one of the three areas can cause a contactor or relay to malfunction.

Contacts

The contacts of a relay or contactor must make good direct contact when energized for the device to function properly. One problem often encountered with contactors and relays is the contacts' inability to make good contact. The contacts can be burned, pitted, or stuck together. A set of contacts that are burned or pitted can cause a voltage drop across the contacts.

There are several methods of checking a set of contacts to determine if they are burned or pitted enough to warrant changing the device. The easiest method is to make a visual inspection. Figure 11.1 shows a contactor with a severely damaged set of contacts. Most contactors have movable covers, which allow easy visual inspection. Most relays are sealed and visual inspection is impossible.

A resistance check can also determine the condition of a set of contacts. The device must be energized to check normally open contacts. Normally closed contacts must be checked with the device deenergized. If the resistance is greater than 1 ohm, the contacts should be considered faulty.

A voltage check can also determine the condition of a set of contacts. When a voltage check is made, the contactor or relay should be energized. To make a voltage check on a set of contacts, take a voltage reading from one side of the contacts to the other, as shown in Figure 11.2. The reading will show how much voltage is being lost. The load must be energized when a voltage test is being performed. If the voltage loss across the contacts exceeds 5% of the line voltage, the contacts are faulty and the contactor or relay should be replaced.

The contact alignment can also cause a contactor or relay to malfunction. Contacts should close directly in line with each other and seat directly in line with good firm contact. The major cause of contact misalignment is a faulty mechanical linkage. If the contacts are out of alignment, the contactor or relay must be rebuilt or replaced.

FIGURE 11.1 Contactor with damaged contacts

FIGURE 11.2 Voltage test of set of contacts; 10 volts AC are lost across contacts.

Coil

The coil of a relay or contactor is used to close the contacts by creating a magnetic field that will pull the plunger into the magnetic field. If the coil of a relay or contactor is faulty, the device will not close the contacts. A contactor or relay coil should be checked for opens, shorts, or a measurable resistance. If a coil is shorted, the resistance will be 0 ohms and the coil should be replaced. An open coil will give a resistance reading of infinity, and this coil should also be replaced. A measurable resistance indicates that the coil is good. Almost any measurable resistance indicates a good coil because of the variance in coil voltages. A shorted contactor coil will cause a transformer to burn out, and the service mechanic should take caution not to allow this to happen.

Mechanical Linkage

The mechanical linkage of a contactor or relay can cause malfunctions in many different forms, such as sticking contacts, contacts that will not close due to excess friction, contacts that do not make good direct contact, and misalignment of contacts. The best method for detecting a faulty mechanical linkage is by visual inspection. On contactors and some relays, this can be done by removing the device and merely looking it over. However, most relays are sealed and inspection is impossible. A sealed relay must be checked by determining if the contacts open and close when the relay coil is energized or deenergized.

A mechanical linkage problem can cause a contactor or relay to stick open or closed or cause misalignment of the contacts. If a contactor or serviceable relay has a mechanical linkage problem, it should be replaced (unless it can be easily repaired). A sticky armature can cause a relay, contactor coil, or transformer to burn out.

11.3 OVERLOADS

Most major loads used in heating, cooling, and refrigeration equipment have some type of overload protection. Overloads are often overlooked as being a problem in the system but they may be faulty. A faulty overload can cause the equipment to run without protection or not operate at all. The high cost of the major loads in a system makes it necessary to protect all major loads.

Fuse

The fuse is the easiest type of overload to check because of its simplicity. A fuse can easily be checked with an ohmmeter in most cases. If a 0-ohm resistance is shown, the fuse is good. No continuity indicates a bad fuse.

A fuse on rare occasions will not completely blow or break but will partially burn out. In this case, the fuse will show 0 ohms but will not allow enough current through it to operate the load. A voltage check across each fuse while power is applied to the load will show a partially burned-out fuse. The voltage check is done by placing the leads of a voltmeter across each end of the fuse, as shown in Figure 11.3. If line voltage is read on the meter, the fuse is bad and should be replaced. If no voltage is read, the fuse is good. This method of checking fuses would not always be accurate on 3 phase circuitry.

Circuit Breaker

The circuit breaker is another type of overload device used by some equipment manufacturers and in many electric panels that are commonly used in the industry today. The circuit breaker is a device that will trip or open on an overload and must be manually reset. The circuit breaker is checked by taking a voltage reading on the load side of the circuit breaker, as shown in

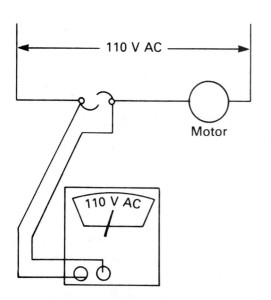

FIGURE 11.3 Voltage check of bad fuse

FIGURE 11.4 Checking a circuit breaker with a voltmeter

Figure 11.4. If line voltage is read on the load side of a circuit breaker, it is probably good.

A circuit breaker can also cause trouble if it trips at an amperage lower than the rated amperage of the breaker. (A clamp-on ammeter is used to check amperage of a circuit breaker.) If this occurs, replace the breaker. A circuit breaker can also cause nuisance trippings if it is unable to handle its rated amperage.

Line Voltage Overload

A line voltage overload installed on a load device is the easiest type of overload to check. It is used on small hermetic compressors and motors and is connected directly to the line voltage supply. The line voltage overload can be open, permanently closed, or open on a lower-than-rated ampere draw. A line voltage overload has only two or three terminals to check. An ohmmeter across the terminals will indicate whether the overload is open or closed.

Care should be taken not to condemn an overload when, in fact, it is open because of a malfunction of the load it is controlling. If a line voltage overload is weak, an amperage check should be made to see what amperage is causing the overload to open. If the dropout amperage is lower than the overload rating, the device should be replaced.

Pilot Duty Overload

A pilot duty overload has a set of contacts that will open if an overload occurs in the line voltage side of the overload. These overloads are arranged so that the line voltage feeds directly through them and then on to the load. The line voltage section of a pilot duty overload can be controlled by heat, amperage, or magnetism—all three are in common use in the industry today. This type of overload is harder to check than the overloads discussed previously because of its complexity. The pilot duty contacts and the controlling line voltage element must be checked.

The pilot duty contacts on a pilot duty overload are easy to check by using an ohmmeter. The pilot contacts are usually easy to distinguish from the line voltage components of the overload because of their small size in relation to the large size of the line voltage connections, as shown in Figure 11.5. The contacts will either be open or closed. If the contacts are open, the overload is bad (or there is an overload in the circuit). If the contacts are closed, the overload is good.

The line voltage part of the pilot duty overload indicates an overload by several methods: heat, current, and magnetism. All three methods determine the current flow, but they use different elements to determine an overload. The heat (thermal) type of pilot duty overload shown in Figure 11.6(a) actually transfers current to heat. The current type of overload is similar in design to the magnetic. The current type of pilot duty overload, shown in Figure 11.5, uses the current through a coil to indicate an overload and open the pilot duty contacts. The magnetic overload shown in Figure 11.6(b) uses the strength of the magnetic field to open and close the pilot duty contacts. The magnetic overload gives a certain amount of built-in time delay due to its makeup.

To check the thermal element of a thermal overload, take a resistance reading across the thermal element. If the resistance is above 0 ohms, the element is bad and the overload should be replaced. The magnetic and

voltage

FIGURE 11.5 Pilot duty and line voltage connections of an overload

(a) Heat (thermal) type of pilot duty overload

(b) Magnetic type of pilot duty overload

FIGURE 11.6 Pilot duty overloads. (Arrows indicate pilot duty connections.)

current types of pilot duty overload can be checked in the same way. However, the magnetic and current overloads use a coil that is larger than the element of a thermal overload. This coil is connected in series with the load and therefore indicates the current being used by the load. The coil in the magnetic or current types of pilot duty overload can be easily checked with an ohmmeter. The resistance of the coil should be 0 ohms because it is part of the conductor going to the load. The overload should be replaced if the coil reads any resistance.

Internal Overloads

Another type of overload is the internal overload used in hermetic compressors. This type of overload is actually embedded in the windings of the hermetic compressor motor, which gives it a faster response to overloads.

The early type of internal overload used separate terminals extending from the inside of the compressor to the outside of the compressor terminal box, as shown in Figure 11.7. This type of internal overload can be simply and easily checked with an ohmmeter to determine if it is open or closed. It is embedded in the winding but makes no electric connections to the windings.

The type of internal overload presently used is hard to check because it has no external connections. It is in series with the common terminal of the compressor motor. This type of overload must be checked as part of the windings of the motor, which makes it extremely hard to diagnose for troubles. The only way to check this type of overload is with an ohmmeter, just as you would check a motor. If an open is present, it could be due to the windings or the overload. Service personnel should never condemn a hermetic compressor that is temporarily overloaded. The compressor should be given ample time to cool.

Many semihermetic compressors have an internal thermostat embedded in the motor windings. The thermostat has a separate connection to the compressor terminals as shown in Figure 11.8. An internal thermostat can easily be checked with an ohmmeter. If any resistance is read, the device is faulty and should be replaced.

Electronic overloads are becoming increasingly popular for the protec-

FIGURE 11.7 Early type of internal overload of a hermetic compressor

FIGURE 11.8 Semihermetic overload terminals

tion of large compressors in the industry. Electronic overloads use sensors with a certain resistance that are placed in the motor windings. The resistances of these sensors change with the temperature of the motor winding: the higher the temperature of the motor winding, the higher the resistances of the sensors. The sensors are connected to an electronic module that amplifies the resistances of the sensors. If the module senses an overloaded condition due to the resistances of the sensors, then the control contacts of the module open and stop the compressor. A more detailed explanation of electronic overloads is given in Chapter 14. Electronic overloads are easy to troubleshoot because most manufacturers give detailed resistance charts on the sensors at certain temperatures. The mechanic only needs to measure the resistances of the sensors to determine if an overloaded condition exists. If an overloaded condition exists, then the pilot duty contacts should be open; but if the motor windings are at normal temperature, the pilot duty contacts should be closed.

11.4 THERMOSTATS

Some type of thermostat is used on most heating, cooling, and refrigeration equipment. Therefore, it is essential to know how to correctly diagnose the condition of thermostats.

There are two basic types of thermostats used in the industry today: the line voltage thermostat and the low-voltage thermostat. The line voltage thermostat is used to make or break line voltage to a load. Its only function is to open or close a set of contacts on a temperature rise or fall. Thus, the line voltage thermostat is usually simpler than the low-voltage thermostat because it does not have as many functions as the low-voltage thermostat. The low-voltage thermostat is used when a voltage lower than 110 volts—usually 24 volts—is used to operate a control system. The low-voltage thermostat can have many functions. It can stop and start a fan motor, operate a fan motor independently of other parts of the system, and do many other functions sometimes required in control systems. The line voltage thermostat is not as accurate as the low-voltage thermostat due to the contacts' larger size, which is necessary to carry the higher voltage. Low-voltage thermostats are usually used on residential heating and cooling control systems and on many commercial and industrial systems. The line voltage thermostat is used on window air conditioners and commercial and industrial air-conditioning, heating, and refrigeration equipment.

Line Voltage Thermostat

The line voltage thermostat is easy to troubleshoot because of its simplicity.

Most line voltage thermostats have two, three, or four terminals, as shown in Figure 11.9. The most important element of checking line voltage thermostats is to be sure the contacts are closed in the correct temperature range. Once it has been determined the thermostat should be opened or closed, it can be checked with an ohmmeter. Or the control voltage can be checked at the equipment. The wires must be removed from the thermostat to check it with an ohmmeter. The ohmmeter will read 0 ohms if the thermostat is closed and infinity if it is open. The service mechanic must determine the correct terminals to check on the thermostat. This information can be found on the unit's wiring diagram. In many cases, it is difficult to remove the wires of a thermostat, so a voltage check is made, as shown in Figure 11.10. When voltage is read across the contacts of a thermostat, the thermostat is open. If no voltage is read, the thermostat is closed. The voltage check must be made with power applied to the unit and the switches placed in an operating position.

Low-Voltage Thermostat

The low-voltage thermostat is more difficult to troubleshoot than the line voltage thermostat because of the many functions of the low-voltage thermostat. The low-voltage thermostat operates the heating and cooling of the system, operates the fan motor with the heating and cooling operations, operates the fan motor independently, often operates two-stage systems,

FIGURE 11.9 Line voltage thermostat

FIGURE 11.10 Voltage check of open thermostat

FIGURE 11.11 Common low-voltage thermostat subbase

operates damper motors, and operates a pilot function of a gas heating system. A common subbase of a low-voltage thermostat is shown in Figure 11.11. All the letter designations, which indicate the different functions of the low-voltage thermostat, are also shown in the figure.

In troubleshooting a system, the low-voltage thermostat and the subbase may be at fault. The low-voltage thermostat and subbase can be checked with an ohmmeter. This can be done at the equipment or at the junction point of the thermostat wires. Service mechanics seldom need to remove the thermostat and install a set of wires on it to check it. A low-voltage thermostat can also be checked by taking a voltage check at the equipment to ensure the thermostat is functioning properly. The low-voltage thermostat and subbase are merely a point in the control system that is fed with low voltage. The thermostat sends a voltage signal to the equipment, which must then operate to meet the conditions called for by the thermostat.

The chart shown in Figure 11.12 can often be used in troubleshooting thermostats.

11.5 PRESSURE SWITCHES

Pressure switches are used on heating, cooling, and refrigeration systems to start or stop some electric load in the system when the pressure in the system dictates this action. Pressure switches are used as safety devices or as operating controls. A pressure switch used as a safety device will stop an electric load when the pressure in a system reaches an unsafe condition. A pressure switch used as a safety device can be used to protect a refrigeration system from excessive discharge pressure or low suction pressure. It can be used in a gas heating system to protect the equipment from low or

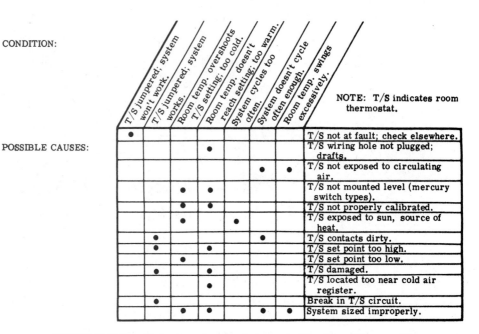

FIGURE 11.12 Troubleshooting chart for thermostats *(Courtesy of Honeywell)*

high gas pressure. It can be used to protect air-moving equipment from low air pressure.

An operating-control pressure switch is used to operate some load in the system. The most common use of an operating-control pressure switch is on a commercial refrigeration system to control the temperature of a walk-in cooler or freezer. Some pressure switches are also used to operate unloading devices on large compressors, to operate condenser fan motors to control the discharge pressure of a refrigeration system, and to operate pumps and cooling tower fans on water-cooled condensers.

The most important aspect of checking a pressure switch is to understand what it is used for in the system. The service technician must also determine if the pressure switch should be opened or closed. Once the service mechanic has determined what the pressure switch is used for and whether it should be opened or closed, it is easy to check the pressure switch. A resistance or voltage check should be made to determine if the pressure switch is opened or closed.

In troubleshooting pressure switches, it may be that the pressure switch is faulty or that the system is malfunctioning. Figure 11.13(a) shows a pressure switch in the control circuit malfunctioning by opening at a higher setting than that on the pressure switch. Figure 11.13(b) shows a pressure switch operating correctly and opening due to a low suction pressure.

The pressure switch can be stuck open or closed, it can open or close on

FIGURE 11.13 Pressure operation

the wrong pressure, or there can be a mechanical problem with the switch itself. If a pressure switch is stuck open or closed, then the pressure in the system will have no effect on the pressure switch. The service mechanic will have no trouble detecting a faulty pressure switch that should be opened or closed and is keeping a load from operating as it should. The mechanic should make sure the pressure switch is not stuck in a position that will result in damaged loads. If a pressure switch is not opening and closing on the right pressure, it should be correctly set or replaced.

On some occasions, a pressure switch has been opened and closed so many times that the switch is worn out. Other mechanical linkage problems include broken springs, leaking bellows, corroded linkages, and broken linkages. Any condition that occurs in the mechanical linkage of a pressure switch will usually call for replacement of the pressure switch.

In most cases when a pressure switch opens, this is due to a malfunction in the system. But occasionally the pressure switch itself is at fault. The service mechanic must determine which condition is occurring. A pressure switch used as an operating control will often have to be reset but seldom replaced. The same troubleshooting procedure is used regardless of the function of the pressure switch.

11.6 TRANSFORMERS

A transformer is a device used to raise or lower the incoming voltage by induction to a more usable voltage for the control system. Some types of transformers are used to buck (lower) or boost (raise) the incoming voltage

FIGURE 11.14 Buck-and-boost transformer

to an air-conditioning unit. A buck-and-boost transformer is used in conjunction with a voltage system that is too high or too low to supply the correct voltage to a system. Figure 11.14 shows a typical buck-and-boost transformer.

A transformer can be checked in two ways: a resistance check or a voltage check. An ohmmeter can be used to check the condition of the windings of a transformer. If the ohmmeter reads 0 ohms, the windings of the transformer are shorted. A reading of infinity indicates an open transformer. A measurable resistance indicates a good transformer. The number of ohms measured would be determined by the voltage of the transformer.

A transformer can also be checked by reading the output voltage if the correct primary voltage is applied. In some cases, the transformer might check out as good if the load is not put across the secondary. But when the load is inserted in the line, the transformer voltage will not be enough to energize the load. In this case, the transformer has a spot burnout and should be replaced. Often transformers are burned out because other devices in the circuit are being shorted. The service mechanic should take every precaution to prevent this from happening.

SUMMARY

Most heating, cooling, and refrigeration personnel are required to do some diagnosing of components. Thus, it is imperative for personnel to understand how components work, to know how to use electric meters to check components, and to know the proper procedures for checking electric components. In this chapter we have presented some guidelines that personnel may find helpful in troubleshooting electric systems and their components.

Contactors and relays can be effectively diagnosed by checking the contacts, the coil, or the mechanical linkage.

Several types of overloads are used in the industry. The fuse can be checked with an ohmmeter. The circuit breaker can often be checked with a voltmeter. The line voltage overload can be checked with an ohmmeter. The pilot duty overload must have its contacts and controlling line voltage elements checked. These components can usually be checked with an ohmmeter. The internal overload used in hermetic compressors can best be checked with an ohmmeter.

Two basic types of thermostats are used in the industry. The line voltage thermostat can be checked with an ohmmeter. The low-voltage thermostat can also be checked with an ohmmeter but care must be taken with this type of thermostat because of its many functions.

There are two types of common pressure switches: low pressure and high pressure. Both can be serviced with either a resistance check or a voltage check. Transformers can also be checked by using either an ohmmeter or a voltmeter.

QUESTIONS

1. What is the first step in troubleshooting any system or component?

2. True or false. The best way to check the contacts of a contactor is by a visual inspection.

3. In troubleshooting contactors and relays, three areas must be checked: the _____, _____, and _____.

4. How do you make a voltage check to diagnose the condition of a set of contacts in a relay or contactor?

5. True or false. If the contacts of a contactor are out of alignment, the contactor must be replaced.

6. If the coil of a relay is shorted, the resistance reading will be ____ ohms.

7. What is the difference between an open coil and a shorted coil?

8. How do you check a fuse?

9. How do you check a circuit breaker?

10. When a line voltage overload is weak, an amperage check should be made to see what _____ is causing the overload to open.

11. What is the difference between a line voltage overload and a pilot duty overload?

12. True or false. The internal overload is difficult to check because it is in parallel to the common terminal of the compressor.

13. The service technician should never condemn a compressor with an internal overload until it has had time

to _____.

14. The two types of thermostats used in the industry are _____ and _____.

15. Give some common applications of the low- and line voltage thermostats.

16. The line voltage thermostat usually has _____ function, while the low-voltage thermostat has _____ functions.

17. What do the letters R, Y, W, and G represent on a thermostat subbase?

18. How do you diagnose a bad pressure switch?

19. The most important aspect of checking a pressure switch is to understand its _____ in the system.

20. What is the proper procedure for checking a transformer?

LAB SECTION REFERENCE

For experiments and activities dealing with material covered in this chapter, refer to Lab Number 8, Transformers; Lab Number 9, Contactors and Relays; Lab Number 10, Thermostats; Lab Number 11, Pressure Switches; and Lab Number 12, Overloads.

12

Air-Conditioning Control Systems

INTRODUCTION

In the heating, cooling, and refrigeration industry there are many different types of control systems and they range from simple to very complex. The equipment used in residential air-conditioning units is usually simple and contains a limited number of components. Commercial and industrial air-conditioning equipment uses a more complex control system, with more emphasis on safety devices than is the case for residential control systems. The trend in the industry in the last few years has been to make residential air-conditioning control systems simpler but without any reduction in the safety. This has been done because of advances in control system design. At the present time the industry is using some solid-state control modules in control systems, but the control systems are by no means becoming completely solid state.

The control systems used in the industry today are combinations of different control systems for separate pieces of equipment, such as a furnace with an air-conditioning unit containing some method of interlocking. The control system has the same function whether the system is a single component or a combination of components. An air-conditioning control system is designed to operate a system automatically, incorporating all the necessary safety equipment by the use of a thermostat. The safety devices of a control system ensure that the loads of the system operate without any chance of damage. However, these safety devices are not always effective in preventing damage to the components because of the economy factor that must always play a part in the design of equipment. Beyond the safety devices of a control system are devices that control the operation of equipment to maintain a temperature at a certain range.

The installation mechanic of air-conditioning or heating equipment must be able to connect the control system of the equipment and the thermostats—and in many cases the furnace. Most manufacturers furnish a wiring diagram with equipment that shows the proper hookup of the equipment being used. In many cases, however, the installation mechanic must install equipment without a diagram. Therefore, it is important for service personnel to know how to connect the controls correctly to ensure proper operation.

All heating, cooling, and refrigeration personnel are at one time or another required to be familiar with modern control systems. For example, salespeople may be called on to assist customers with control system design. Engineers are required to design control systems. Service mechanics are required to maintain and repair control systems.

About 85% of the problems in heating, cooling, and refrigeration control systems can be traced to some electrical problem. Although it is impossible to study all the different control systems used in the industry today, familiarity with the more common control systems will enable you to understand special control systems because of the similarities of design. In this chapter we will present the more common control systems in use in the industry today.

12.1 BASIC CONDENSING UNITS

A **condensing unit** is the portion of a split air-conditioning system that is mounted outside and contains the compressor, the condenser, the condenser fan motor, and the necessary devices to control these components. A split system is one that is divided into two parts, usually a condensing unit (outside) with a fan coil unit (inside). Figure 12.1 shows an air-cooled condensing unit.

In most cases, the condensing unit is used with some type of equipment that will produce the air flow, and a coil must be mounted in the air flow. Most condensing units used in the industry today are air-cooled condensing units, which means that the condenser is cooled by air. Water-cooled condensing units may still be found in the field, but these units are rapidly being replaced with the smaller and more economical air-cooled condensing units.

When a condensing unit is used in an air-conditioning system, it is separate from the rest of the equipment. The control system must be connected to allow for complete control of the entire system.

The condensing unit takes the cool suction gas coming from the evaporator, compresses it, condenses it from a gas to a liquid, and forces it back to the evaporator. The standard components of an air-cooled condensing unit are

FIGURE 12.1 Air-cooled condensing unit

FIGURE 12.2 Components of an air-cooled condensing unit

the compressor, the condenser, the condenser fan motor, and all the devices used for control. The components of an air-cooled condensing unit are shown in Figure 12.2. Most of the controls used to operate the condensing unit are mounted somewhere in the unit itself with the exception of the interconnection between the thermostat and the condensing unit.

Condensing Unit Components for a Simple Control System

In early control systems all components necessary for operation of the total air-conditioning system were mounted in the condensing unit. This practice has all but disappeared in the industry because of the expense that it required and the realization that it was unnecessary. Most modern condensing units pick up their 24-volt power supply to operate the control system from the furnace transformer or from an indoor fan relay package. The evaporator fan motor usually is controlled by a relay when the system contains a furnace or fan coil unit and the condensing unit.

The compressor and condenser fan motor in a condensing unit are usually controlled by a contactor. The simplest control system used on condensing units today is a contactor that controls the operation of the entire condensing unit with the exception of the internal overloads in the compressor and condenser fan motor. With this type of control system, the contactor is energized whenever the thermostat is calling for cooling, even though one of the safety devices has the compressor or condenser fan motor cut out. This control system is the simplest and the least expensive of any condensing unit control system used today. A separate 24-volt source must be supplied from other components through the thermostat to the condensing unit.

All condensing units without a high-pressure switch must have an internal **relief valve** in the compressor that opens if the discharge pressure exceeds an unsafe level. This relief valve replaces the high-pressure switch on some units.

Figure 12.3 shows a control panel used on a condensing unit control system. Note the simplicity of the panel.

Figure 12.4 shows the schematic diagram of this type of control system. The control system requires a supply of 24 volts to the contactor through the thermostat to operate the condensing unit. When 24 volts are applied to the contactor through the thermostat, the contactor will close, allowing 230 volts to be applied to the compressor and condenser fan motor, and they will start. This control system is easily diagnosed for trouble, with the exception of the overloads in the compressor and the condenser fan motors.

Condensing Unit Components for Complex Control Systems

Many manufacturers use a complex control system that contains certain safety devices to ensure safe operation of the condensing unit. The control

FIGURE 12.3 Control panel of a simple air-cooled condensing unit

FIGURE 12.4 Schematic diagram of a simple air-cooled condensing unit with a minimum number of controls

system discussed in the preceding section had only the minimum required safety devices, which are often inadequate for a completely safe operation of the condensing unit. The more complex control systems incorporate a high-pressure and a low-pressure switch. Figure 12.5 shows the schematic diagram of a control system with the extra safety controls. If any of the safety controls open, the compressor and condenser fan motor stop.

Several manufacturers also use a device that protects the system from **short cycling**. This device maintains a certain period of time between the cycles of the equipment so the system does not cut on and off in rapid succession (short cycling). This control system adds a relay and a time clock mechanism for this purpose. Other systems use a special relay to lock out the entire condensing unit until it can be reset manually.

There are many other control system designs that use more components for specific purposes. For example, large condensing units use a control relay or compressor relay to control a 240-volt contactor coil and the condenser fan motor. This arrangement is used because for a large contactor, 24 volts are less effective in pulling the contactor in than are 240 volts. Many other designs, too numerous to discuss, are used in the industry for the control of condensing units. In most cases, though, control systems have

Legend

C:	Contactor		CRC:	Compressor Running Capacitor
COMP:	Compressor		HP:	High-Pressure Switch
CFM:	Condenser Fan Motor		LP:	Low-Pressure Switch
FRC:	Fan Running Capacitor		CST:	Compressor Safety Thermostat

FIGURE 12.5 Schematic diagram of a condensing unit with safety controls

several basic similarities that make them fairly easy to understand.

Wiring

All condensing units come from the manufacturers with a wiring diagram and, in some cases, with an installation wiring diagram. Usually it is not difficult to follow the installation instructions and wire the condensing unit correctly. Figure 12.6 shows an installation diagram used to install a condensing unit to a furnace for an add-on air-conditioning system. This diagram shows the wiring hookup of a condensing unit and a gas furnace. It is used when installing a condensing unit and furnace.

Heating-Cooling Thermostat Model

NOTES:

1. Be sure power supply agrees with equipment nameplate(s).
2. Low voltage (24 V) wiring to be No. 10 A.W.G. min.
3. Grounding of equipment must comply with local codes.
4. Maximum external load shall not exceed 28 VA.

FIGURE 12.6 Installation diagram for a condensing unit connected to a furnace *(Courtesy of General Electric Co., Central Air Conditioning Dept.)*

Troubleshooting

Any troubleshooting of the condensing unit can be done from the schematic because of the simplicity of most control systems. By using the schematic diagram, and having an understanding of the components, you should have no trouble with diagnosing a residential control system. However, you may have to study the more complex commercial and industrial control systems to diagnose problems in them.

12.2 PACKAGED UNITS

A **packaged air-conditioning unit** is built with all the components housed in one unit. In most cases, packaged units are complete except for the power connections and the control connections.

Air-cooled packaged units are the most widely used types of packaged units. The units are used on applications ranging from small residential cooling units to large commercial and industrial air-cooled units. The smaller residential systems may be installed when the structure is built or at a later date.

An air-cooled packaged unit would usually be mounted outside the structure, as shown in Figure 12.7. In some cases, air-cooled packaged units can be installed inside with a remote condenser. The air-cooled packaged unit contains the compressor, evaporator fan motor, condenser fan motor if not of remote design, and all the necessary controls.

Water-cooled packaged units are usually used on commercial and industrial systems. The water-cooled packaged unit is shown in Figure 12.8. They are usually installed when the structure is built. The water-cooled unit requires a cooling tower or some means of supplying water to the condenser. If a

FIGURE 12.7 Air-cooled packaged unit   FIGURE 12.8 Water-cooled packaged unit

water-cooled unit is used, there must be interlocks to ensure that the cooling tower pump is operating when the compressor is operating. Water-cooled packaged units come complete with the compressor, evaporator fan motor, all the necessary controls, and the condenser.

Control System for Packaged Air-Conditioning Systems

Packaged unit controls are usually simple on small residential equipment and become more complex on the larger commercial and industrial equipment. The smaller air-cooled packaged units use a fairly simple control system with an interlock for the condenser fan motors. The large air-cooled and water-cooled units have a fairly complex control system. The large air-cooled packaged units are complex because of the zoning requirements that these units must meet in the rooftop applications in which they are usually used. Water-cooled packaged units utilize a line voltage control system with some type of interlock to start the cooling tower pump or other accessories that are required.

Air-cooled packaged units can incorporate some form of heating as well as cooling. The smaller packaged units usually have some easy means of installing electric resistance heat into the equipment. If gas heating is desired, a **gas pack** should be used, which is an electric air-conditioning and gas-heating system mounted in one unit. The larger air-cooled packaged units make electric and gas heat available to the consumer. Water-cooled packaged units and air-cooled packaged units usually use a hot-water coil, steam coil, or resistance heat for heating purposes.

Small Air-Cooled Packaged Units

The small air-cooled packaged units usually have a simple and easy-to-follow control system. The control system uses a contactor to operate the compressor and the condenser fan motor and an evaporator fan relay to control the evaporator fan motor. All necessary safety components are also included in the control system.

All manufacturers' control systems are not alike. Many use only the necessary safety controls, while others use more safety controls to ensure complete and safe operation.

Figure 12.9 shows the schematic diagram of a small air-cooled packaged unit and its controls. This unit would be used in a residence or small commercial application. The thermostat at the bottom of the schematic controls the operation of the unit. When the fan switch is in the "on" position or in the "auto" position, the thermostat calls for cooling, energizes the indoor fan relay, and, through its contacts, starts the indoor fan motor. The compressor is energized through a number of components. First the thermostat closes, energizing the control relay and closing its contacts. This energizes the outdoor fan motor and the contactor coil through the holding relay and the timer circuits. Once the contactor closes, the compressor will begin operation.

The installation of the air-cooled packaged unit is relatively simple because

Legend

TRANS:	Transformer
TM:	Timer Motor
IFR:	Indoor Fan Relay
HR:	Holding Relay
CR:	Control Relay
OFMC:	Outdoor Fan Motor Contact
HP:	High-Pressure Switch
LP:	Low-Pressure Switch
C:	Contactor
IFM:	Indoor Fan Motor
MS:	Momentary Switch
OFM:	Outdoor Fan Motor
HTR:	Crankcase Heater
IT:	Internal Thermostat
OL:	Overload
RC:	Running Capacitor
SC:	Starting Capacitor
SR:	Starting Relay

FIGURE 12.9 Schematic diagram of a small air-cooled packaged unit used in a residence *(Reproduced by permission of Carrier Corporation. Copyright © 1977 Carrier Corporation, Syracuse, NY)*

the only necessary connections are from the power source and the control source or thermostat. Most manufacturers include an installation diagram giving the correct connections and the control layout of the system.

Gas-Electric Air Conditioners

The combination of electric air conditioning and gas heating in a packaged unit gives a system that is somewhat more complex than a straight air-cooled packaged unit. The main difference between the two units is all the extra components needed to combine a heating system with an air-conditioning system. The basic control system for the air conditioning is the same as the one for most small residential units. However, several components are added to take care of the heating control system.

Figure 12.10 shows a schematic diagram of the control system used on a modern gas-heating and electric air-conditioning packaged unit. The installation instructions should be followed when installing this type of system.

Referring to Figure 12.10, the cooling mode of operation is energized through the thermostat to the contactor, which starts the compressor and outdoor fan motor. The indoor fan motor is started through the thermostat to the blower relay. The heating mode of operation is by the first stage of the thermostat (heating) energizing the gas valve.

Air-Cooled Packaged Units with Remote Condensers and Water-Cooled Packaged Units

The air-cooled packaged unit with the remote condenser and the water-cooled packaged unit use a line voltage control system and are similar in circuitry. These systems usually have line voltage control because they are shipped from the factory with all the controls mounted and wired, including the thermostat. These units are merely set in place when installed. The only necessary installation would be the power wiring connection, the condensate drain, and the piping.

The major difference between the water-cooled unit and the air-cooled unit is in the method the manufacturer uses to interlock the necessary components to cool the condenser. Figure 12.11 shows the schematic diagram of an air-cooled packaged unit with a remote condenser interlocked in the system. Figure 12.12 shows the schematic of a water-cooled packaged unit with a cooling tower pump interlock.

The unit shown in the schematic diagram of Figure 12.11 has a line voltage control circuit. The switch marked "off," "fan," and "cool" operates the unit. With the switch set on "fan," the only portion of the unit that can

FIGURE 12.10 Schematic diagram for electric air conditioning and gas heating unit *(Courtesy of Rheen Air Conditioning Division Fort Smith, AR)*

SCHEMATIC

FIGURE 12.11 Schematic diagram of an air-cooled packaged unit with a remote condenser *(Reproduced by permission of Carrier Corporation. Copyright © 1977 Carrier Corporation, Syracuse, NY)*

FIGURE 12.12 Schematic diagram of a water-cooled packaged unit *(Reproduced by permission of Carrier Corporation. Copyright © 1977 Carrier Corporation, Syracuse, NY)*

operate is the indoor fan motor; it is energized through the manual switch. If cooling is desired, the switch must be set on "cool," which will not interrupt the operation of the indoor fan motor. This circuit energizes the outdoor fan motor through the outdoor fan relay. The compressor is energized through the contacts of the contactor, which is energized through the holding relay and the timer.

The unit shown in the schematic diagram of Figure 12.12, a small water-cooled packaged unit, also has a line voltage control circuit. The switch marked "fan" and "cool" actually stops and starts the unit. When the switch is in the "fan" position, the fan will run. When the switch is moved to "cool," the fan will continue to run and the cooling thermostat will operate the compressor through the contactor.

Installation of these air conditioners is simple because no connections are necessary, with the exception of the power wires. However, water connections are necessary on the water-cooled units and refrigerant lines are needed between the outdoor remote condenser and the inside packaged unit.

Rooftop Units

The air-cooled packaged units used in rooftop applications are complex because they usually have some type of heating to go along with the air conditioning and, in some cases, some type of zone control. The air-conditioning and heating control system is usually simple. However, when it is connected to all the zone controls, the system becomes difficult to service and install.

As their name implies, these units are usually installed on the roof of a structure. But they could just as easily be installed on ground level when the necessary changes are made in the duct work connections. Often the electric connections of these systems are hard to install because of the number of wires required for the control circuit and the power wiring. The installation instructions furnished with these units are usually well written and give explicit instructions.

12.3 FIELD WIRING

In all heating, cooling, and refrigeration electric systems there is a certain amount of wiring that must be connected to the equipment after it has been set in place. This wiring must be installed by the installation mechanic and is called **field wiring**.

The **factory-installed wiring** is the wiring installed at the factory. It usually takes care of the connections between the components in the control panel and the system components. The factory wiring has been sized, color-coded, and installed in the control system to operate the equipment properly. The remainder of the wiring, whether it be the power wiring or control wiring, must be connected in the field.

Power Wiring

The power wiring of an air-conditioning and heating system is usually simple and easy to install on residential systems. There are two power connections that must be made on a split air-conditioning and heating system: the connections to the condensing unit and the connections to the evaporator fan motor or furnace. Figure 12.13 shows the power wiring on a condensing unit and a furnace with the correct connections from the distribution panel in the residence.

The power connections for a residential air-cooled packaged unit are shown in Figure 12.14. This circuit is relatively simple, containing only the power connections and some type of bonding ground.

The electric supply wiring of a commercial and industrial system is some-

FIGURE 12.13 Power wiring for a condensing unit and a furnace. (No control wiring is shown.)

FIGURE 12.14 Power connections for a small air-cooled packaged unit. (No control wiring is shown.)

what more complex than the simple residential systems. The use of three-phase current in commercial and industrial structures does not add to the complexity of the system, only to the number of power wires that must be supplied to the system. The evaporator fans on any split commercial and industrial system would be supplied with a separate power source. The condensing unit would be supplied with its own power source. Figure 12.15 shows the supply wiring that must be installed with a fan coil unit that uses three-phase current.

Heating Systems.   Commercial and industrial systems use various types of heating systems. In a system that uses electric resistance heat, a large power supply would be required to operate the heat. If some type of gas heating is used, the wiring would not have to be as large as that used with electric heat. Figure 12.16 shows the supply wiring hookup of a condensing unit, fan coil unit, and an electric duct heater. In many cases hot water or steam is used, but there are no additional power connections that must be made except for the pumps.

Large rooftop or packaged units equipped with gas heating usually require only one power source to the entire system. The rooftop units that incorporate electric resistance heat are usually supplied by two or more power sources because of the large load created by the electric heat. One of the power sources feeds the air-conditioning components and the evaporator fan motor; the other source feeds the electric heat.

FIGURE 12.15 Power wiring for a three-phase fan coil unit. (No control wiring is shown.)

There are many types and manufacturers of equipment in the industry today and it is impossible to show all the supply power connections in this section. However, it is important that the installation mechanic study other methods of connecting the power wiring to modern equipment.

Sizing Wires and Fuses. When installing heating, cooling, and refrigeration systems, air-conditioning personnel in charge of the installation must make sure that all power wiring is of the correct size and type. The distance that the power wiring must be run is important because of the voltage drop that can occur on long circuit runs. The installation instructions usually give a wiring chart with the length that is allowable for each size of wire. Installation personnel should be capable of sizing the wire if it is not listed in the installation instructions (refer again to Chapter 6).

It is also important to follow the manufacturer's recommendations for the fuse or breaker size. Remember: Before any heating, cooling, or refrigeration system can be expected to operate properly, it must first be supplied with the correct size of wire and fuse to deliver the proper voltage to the system.

Control Wiring

The control wiring of heating and air-conditioning systems is just as important as the supply power wiring. In most cases, low-voltage control systems are used on residential and small commercial applications. On large com-

Legend

IFMS: Indoor Fan Motor Starter
IFM: Indoor Fan Motor
CFM: Condenser Fan Motor
COMP: Compressor
CFM: Condenser Fan Motor Contactor
C: Contactor

FIGURE 12.16 Supply power wiring for a condensing unit, fan coil unit, and large electric duct heater. (No control wiring is shown.)

mercial and industrial applications, line voltage control systems are often used. Many large commercial and industrial systems are controlled by special control systems. These specialized control systems can be electric, pneumatic, or electronic. Specialized control systems are a complete subject in themselves and hence will not be covered in this section.

Residential. All residential packaged air conditioners use a low-voltage control system that connects the thermostat to the unit, as shown in Figure 12.17. These control systems are simple and easy to install by following the installation instructions.

Residential split systems are usually easy to install. They all incorporate a 24-volt control system, with the low-voltage power being supplied from an

Power supply
208–230/60/1

L_1 L_2

Power supply
208–240/60/3

L_1 L_2 L_3

Disconnect
switch
(not
provided)

Black L_3 △2
Black L_2
Black L_1
Black
Black △1

Equipment
ground

Blue   R
Red   Y1
White   G
Yellow △1

R
Y
G
W

Room thermostat

Low-voltage
junction box

△1 These leads are used only when an electric heater is installed.

△2 Do not cut excess length from these leads if an electric heater
will be installed at a later date.

FIGURE 12.17 Control connections for a residential packaged unit *(Courtesy of Lennox Industries, Inc.)*

FIGURE 12.18 A common indoor fan relay package

indoor fan relay package, the furnace, or the condensing units. Figure 12.18 shows one type of indoor fan relay package that is in common use in the industry today. The **indoor fan relay package** acts as a junction point between the thermostat, the furnace, and the condensing unit. The package has a terminal board labeled with easily identifiable letter identifications.

The control system of a residential split system can be connected several different ways, and the manufacturer's installation instructions should be followed. Some manufacturers use a transformer mounted in the condensing unit that supplies the low-voltage to the system, but this practice is becoming unpopular because of the advancements made in the indoor fan relay packages. In some cases, the furnace transformer is used as the low-voltage power supply, but caution should be taken to ensure that the proper size is used. The more modern condensing units are usually designed without the low-voltage transformer but with two low-voltage connections instead. Figure 12.19 shows a control layout for a furnace, indoor fan relay package, thermostat, and the condensing unit. Figure 12.20 shows a control system of a furnace and a condensing unit with the thermostat and the low-voltage power supply coming from the furnace transformer.

FIGURE 12.19 Control layout for a furnace, indoor fan relay package, thermostat, and condensing unit *(Reproduced by permission of Carrier Corporation. Copyright © 1977 Carrier Corporation, Syracuse, NY)*

FIGURE 12.20 Control system of a furnace and a condensing unit with the thermostat and low-voltage power supplied from the furnace *(Reproduced by permission of Carrier Corporation. Copyright © 1977 Carrier Corporation, Syracuse, NY)*

Industrial. Control systems for relatively small commercial units are similar in design to the residential low-voltage system. The only exception is that the system uses a control relay that supplies the condensing unit with 230 volts while using a normal 24-volt control system. Two-stage heating or two-stage cooling can also be used with a low-voltage control system.

Sizing Wire. Control wiring in most cases is a small wire (No. 18 to No. 20) that is usually covered by a rubber jacket when using a low-voltage control system. Thermostat wire can be purchased single-stranded or multi-stranded, with or without a rubber jacket protecting the small thermostat wires. The size of the control wiring should follow the manufacturer's specifications.

The wiring of air-conditioning and heating systems is one of the most important factors in the installation of equipment. The control system is actually the heart of the electric system because its function is to properly control the entire system. The life of the equipment can be drastically cut by undersized wire or by having loose connections. Therefore, wiring to all components must be sized correctly. The installation instructions are the best place to look for the proper method of wiring and the proper connection points. Overlooking the smallest detail in the electric connections of a system can give trouble and improper operation.

SUMMARY

Air-conditioning, heating, and refrigeration control systems range from simple control systems with a simple line voltage thermostat to very complex systems using many different devices to control the temperature of each zone in a large structure. It is understandable that a large air-conditioning unit with a very expensive compressor will require a larger and better control system than the smaller residential units.

The electric system of an air-conditioning unit covers the power wiring as well as the control wiring, and in most cases they should be treated as one.

The electric system of a small condensing unit used in the residential or small commercial area is designed with only the bare necessities in controls. Most of these smaller units are designed with a contactor and few safety controls in the condensing unit. The 24-volt supply must come from other equipment. The compressors of these units almost always contain an internal overload and this should be taken into consideration when working with these units. Some of the smaller commercial systems will have various added features to prevent the equipment from short cycling and to provide other important advantages.

Packaged air-conditioning equipment can take two different forms. The small residential and commercial packaged units are completely self-contained and must have an outside air source. The straight commercial packaged units are built in one piece and are used in many cases in a free-standing position within a structure; they may be water-cooled or air-cooled with a remote condenser.

The packaged units that are mounted outside come in various designs. They range from a 2-ton air conditioner for a residence to a 50-ton or larger rooftop unit used in commercial buildings.

The straight packaged units usually use a low-voltage control system. Some of the larger units will have a line voltage control system within the unit. The free-standing packaged units almost always have a line voltage control system because they are prewired at the factory in this manner. All packaged units are prewired at the factory. In the straight packaged units, the control wiring must be field installed. The free-standing packaged units are completely wired at the factory.

All heating, cooling, and refrigeration equipment will require some kind of electric connection in the field, if nothing more than the power connections. Most equipment will require a control circuit and a power circuit before the equipment will operate. The power circuit consists of the electric source carried to the unit. In some cases, the evaporator fan motor will require a separate source. Control wiring is required between the thermostat and the equipment, or two sections of equipment, to make one complete system. The manufacturer almost always supplies an installation diagram with the equipment. This diagram will give wire and fuse sizes that are recommended by the manufacturer and should be followed.

QUESTIONS

1. What is the purpose of an air-conditioning control system?

2. What is the major difference between the controls on a small residential condensing unit and a large commercial condensing unit?

3. What is a condensing unit?

4. Briefly describe how a condensing unit operates.

5. Most condensing units pick up the 24-volt power supply to operate the control system from _____ or _____.

6. True or false. The simplest control system used on condensing units is a contactor.

7. An internal relief valve is used in a compressor in place of a _____.

8. Explain the operation of a device that prevents short cycling in a system.

9. What is a split air-conditioning system?

10. Draw a diagram of the wiring of a simple residential air-cooled condensing unit used with a furnace, having a contactor, condenser fan motor, and a compressor.

11. What is the usual control voltage of a residential heating and cooling system and from where does it receive it?

12. How would a low-pressure and a high-pressure switch be connected into the electric system of a residential system?

13. What would be the reason for using a 240-volt control system in a large condensing unit?

14. What is a packaged air-conditioning unit?

15. An air-cooled packaged unit is usually mounted _____.

16. Why is it necessary to interlock the cooling tower pump and fan motor into the control circuit of a water-cooled packaged unit?

17. A gas pack is _____.

18. Line voltage control systems are used with what types of packaged units?

19. What is the major difference between the water-cooled unit and the air-cooled unit?

20. What is field wiring?

21. Name the two types of power connections that must be made on a split air-conditioning and heating system.

22. True or false. Commercial and industrial systems usually use electric resistance heat.

23. What are two major considerations an installation mechanic must take into account when installing air-conditioning power wiring?

24. What type of control system is used on most residential and small commercial equipment?

25. What is the purpose of an indoor fan relay package?

26. Control wiring in most cases is _____.

27. The heart of the electric system in heating, cooling, and refrigeration equipment is _____.

28. What type of electric connections must be made to a residential packaged unit by the installation mechanic to ensure proper operation?

29. What type of electric connections must be made to a commercial and industrial packaged unit by the installation mechanic to ensure proper operation?

LAB SECTION REFERENCE

For experiments and activities dealing with material covered in this chapter, refer to Lab Number 13, Wiring A Split Air-Conditioning System, and Lab Number 14, Wiring A Packaged Air-Conditioning Unit.

13

Control Systems: Circuitry and Troubleshooting

INTRODUCTION

Control systems used in the heating, cooling, and refrigeration industry are designed to control the electric loads of a system to maintain the desired temperature of a given area or medium. Control systems must contain the necessary safety components to ensure safe operation of the equipment.

Many different types of control systems are used in the industry today, but most systems have certain components and circuits in common, depending on the control voltage and the type of equipment. Low-voltage control circuits are different from line voltage control circuits because of the devices that can be used with each type of system. The basic control system contains several components common to all control systems, such as the compressor, the evaporator fan motor, and the condenser fan motor, along with their controlling devices.

It is essential that installation and service mechanics understand the operation of the basic circuits of control systems to be able to install or troubleshoot the smaller air-conditioning and heating systems that are common to the industry. All equipment will not have the same type of controls and may use different methods of controlling certain devices. In this chapter we will be looking at basic control system circuits and procedures for troubleshooting control systems.

13.1 BASIC CONTROL CIRCUITS

All control systems have certain circuits and components in common. Most air-conditioning systems have a contactor to start and stop the compressor. The compressor is the largest electric load in an air-conditioning system. Therefore, it is necessary to use a contactor or starter to control its ampacity.

The control of the condenser fan motor is common on many systems, but it does vary due to the new high-efficiency equipment that is being produced. Circuits controlling the evaporator fan motor are common because of the use of an evaporator fan relay, but on heating systems different methods are used. The safety devices used on control systems vary greatly, but they are usually connected into the control circuit in series.

Now let us look at these control system components in detail.

Compressor Control Circuits

The first basic circuit of almost any residential or small commercial air-conditioning control system is the device that starts and stops the compressor. On a low-voltage control system, the thermostat opens or closes to energize or deenergize the contactor, which starts or stops the compressor, as shown in Figure 13.1. This circuit is common to almost all residential and small commercial air-conditioning systems in operation today.

The condenser fan motor is usually connected directly to the contactor to ensure that it is operating whenever the compressor is in operation, as shown in Figure 13.1.

Evaporator Fan Motor Control Circuits

All evaporator fan motor circuits are almost alike in the smaller ranges of equipment. The evaporator fan motor operates from a relay controlled by the thermostat. If the thermostat fan switch is set on the "on" position, the indoor fan relay will be energized by the cooling function of the thermostat. Figure 13.2 shows the indoor fan relay with the connections of the fan motor and the thermostat. Notice the fan switch on the thermostat and note how the indoor fan is cut on or off by use of the indoor fan relay. This control system is common only to air-conditioning systems. On heating installations, other controls are needed to start the fan.

Some very small packaged units use a unique control system for the evaporator fan motor, as shown in Figure 13.3. This type of control of the evaporator fan motor is used because there is only one fan motor to operate, which works as both an evaporator fan motor and a condenser fan motor. In this case, it is imperative the fan motor operate whenever the compressor operates in order to cool the condenser. When the fan is to be energized on the thermostat through the fan switch for continuous opera-

Legend

C:	Contactor
COMP:	Compressor
CFM:	Condenser Fan Motor
CRC:	Compressor Running Capacitor
CFMC:	Condenser Fan Motor Capacitor
TR:	Transformer

FIGURE 13.1 Schematic diagram of a contactor circuit controlling a compressor, showing the connection of the condenser fan motor

tion, the fan is energized through an indoor fan relay. When the fan is energized through the cooling sections of the thermostat, the fan is controlled by the contactor. Most evaporator fan motor circuits are controlled by an indoor fan relay. Only in rare instances will the evaporator fan motor be energized by some other means.

Condenser Fan Motor Control Circuits

Most condenser fan motors on residential air-conditioning systems are controlled by the contactor because of the simplicity and the economy of using

Thermostat

FIGURE 13.2 Schematic diagram of the connection for the indoor fan motor and its control

one control. Some condenser fan motors are cycled on and off through an outdoor thermostat set at a temperature that will maintain a constant head pressure. This allows better system efficiency. This type of control is accomplished on residential and small commercial systems by inserting the thermostat in series with the condenser fan motor. Other methods are used in the larger commercial and industrial applications.

Safety Control Circuits

Most safety controls used in residential and small commercial systems are

Legend

——— Factory Wiring
– – – – Optional Starting Capacitor, and Relay Wiring
━━━ Field Wiring
AHA: Adjustable Heat Anticipator
C: Contactor
CC: Cooling Compensator
CH: Crankcase Heater
COMP: Compressor Motor
FM: Fan Motor
FR: Fan Relay
HP: High-Pressure Switch
LP: Low-Pressure Switch
RC: Running Capacitor (COMP)
SC: Starting Capacitor (COMP)
SR: Starting Relay (COMP)
TC: Cooling Thermostat
TH: Heating Thermostat
TR: Transformer

NOTES:

1. Compressor motor and fan motor provided with inherent thermal protector.
2. Dual element fuse size 30 A max.
3. Terminals Y, G, R are for connection of N.E.C. class 2 control circuits at 24 V max — external load 15 VA between R and C.
4. On the fan motor the black lead is for high speed, the red for low speed, the yellow lead is common.

FIGURE 13.3 Schematic diagram of a small packaged unit with one fan motor *(Reproduced by permission of Carrier Corporation. Copyright © 1977 Carrier Corporation, Syracuse, NY)*

connected in series with the contactor coil. The only exception is the internal thermostat in a compressor, which breaks the power wires inside the compressor. Any other safety devices used in the control system, such as a high-pressure switch or a low-pressure switch, are connected in series with the contactor coil. The series connection ensures that in the case of an unsafe condition, the compressor will be deenergized.

Figure 13.4 shows a high-pressure and a low-pressure switch in the control circuit. Any additional safety devices would be connected in the same manner. The internal overload in the compressor is connected in series with the common terminal of the compressor so the compressor will be deenergized when an overload occurs.

Legend

C: Contactor
COMP: Compressor
CRC: Compressor Running Capacitor
CFM: Condenser Fan Motor
CFMC: Condenser Fan Motor Capacitor
IFR: Indoor Fan Relay
IFM: Indoor Fan Motor
HP: High-Pressure Switch
LP: Low-Pressure Switch
CH: Crankcase Heater

FIGURE 13.4 Total control circuit of a residential or small commercial air-conditioning system

13.2 TOTAL CONTROL SYSTEM OF RESIDENTIAL UNITS

The total control circuit of a residential or small commercial air-conditioning system is shown in Figure 13.4. Each circuit can be discussed separately as long as each part in the circuit is complete in the total operation of the control system.

The thermostat is a switching point of the low-voltage supply to the necessary components. Figure 13.5 shows the schematic of the thermostat. From the schematic it can be seen that the indoor fan motor can operate continuously or cycle with the cooling thermostat, depending on the position of the fan function switch. The remainder of the thermostat oper-

FIGURE 13.5 Schematic of a thermostat for a total residential control system *(Courtesy of Honeywell)*

ates the heating or cooling or both. The cooling thermostat will close or open to the correct temperature, as well as control the heating by the heating thermostat.

Figure 13.3 shows that the compressor and in most cases the condenser fan motor operate when the contactor is energized. The contactor is energized by the cooling thermostat until the temperature is decreased to the desired range. The evaporator fan motor is energized on cooling by the indoor fan relay and on the heating cycle by a fan switch contained in the furnace.

As can be seen in Figure 13.4, the safety devices are connected in series with the contactor coil, with the exception of the internal overload in the compressor, which is in series with the compressor common terminal. In some cases, manufacturers use a crankcase heater that is energized at all times to keep liquid refrigerant from migrating into the compressor crankcase. This control system is commonly used on residential and small commercial air-conditioning and heating systems.

13.3 ADVANCED CONTROL SYSTEMS

All types and designs of control systems are used in commercial and industrial air-conditioning, heating, or refrigeration systems. The control systems used in equipment from 7.5 to 100 tons cannot afford to be as simple and use as few components as the smaller systems. This type of control system must ensure the electric loads are operating in a safe manner because of their greater cost. The control of this type of equipment must also be more complex than the smaller systems because of its size and operation.

Most large systems use some method of capacity control because of the variations in the load in the conditioned area. When capacity control is used on these larger air-conditioning systems, it usually involves a more complex control system. Also, most large air-conditioning and heating systems have several important control circuits that prevent short cycling and operation of the equipment under conditions that would cause damage. Thus, we see that large equipment uses many more components than small equipment because of size, cost, capacity control, and general design of the larger systems.

Compressor Motor Controls

It would be impossible to cover all the control circuits used in the large equipment. However, the more common control circuits will be covered in detail.

There are two methods of energizing the compressor on the large systems, part winding and across the line. The contactor or magnetic starter is used to energize the compressor on all systems. An across-the-line type of motor starting requires only a single contactor, as shown in Figure 13.6(a). However, a part winding motor would require two contactors along with a time-delay relay. The part winding motor is used to allow the compressor to start easier, with less wear and tear. A part winding motor hookup is shown in Figure 13.6(b). The part winding motor is actually broken down into two separate windings, with the first winding energized 1.5 to 3 seconds before the second winding is energized. This increases the motor's starting efficiency.

Water Chiller Control Systems

Water chillers are refrigeration systems that cool the water pumped into other parts of the system to maintain the desired condition of a specific area. The control on a water chiller will usually come from the chilled water temperature, which will relate the information back to the refrigeration control system. This type of control can also include a pump-down system. The system maintains a certain chilled water temperature at all times. A typical control system for a water chiller is shown in Figure 13.7.

Evaporator Fan Motor Controls

The control of the evaporator fan motor in large commercial and industrial air-conditioning or heating systems comes from a source separate from the

(a) Across-the-line starter (shaded area)

(b) Part winding motor hookup (shaded area)

FIGURE 13.6 Schematic of a part winding motor hookup (shaded areas) *(Reproduced by permission of Carrier Corporation. Copyright © 1977 Carrier Corporation, Syracuse, NY)*

Legend

C:	Compressor Motor Contactor
CCH:	Crankcase Heater
CWI:	Chilled Water Pump Interlock
F_1:	Control Fuse
G:	Ground
HP:	High-Pressure Cutout
HR:	Holding Relay
IT:	Internal Compressor Thermostat
LLS:	Liquid Line Solenoid
LP:	Low-Pressure Cutout
LTC:	Low Water Temperature Cutout
OL:	Overload Relay
PT:	Transformer
R:	Condenser Fan or Pump Motor Relay
SW:	Switch on-off
TM:	Timer Motor
U:	Unloader Solenoid
WTC:	Water Temperature Control
----	Field Wiring
☐	Connections on Terminal Board

NOTES:

1. Interlock between terminals C and D: Condenser Water Pump or Fan Motor Overloads.

2. Remove wire number 9 between terminals C and D when interlock is added.

FIGURE 13.7 Typical control system for a water chiller *(Reproduced by permission of Carrier Corporation. Copyright © 1977 Carrier Corporation, Syracuse, NY)*

condensing unit. Fan motors are generally designed to operate at all times in the larger systems and are cut on and off by a clock or a system switch. These fan motors are interlocked into the control system of the condensing unit to ensure the operation of the fan motor whenever the condensing unit is operating.

The smaller commercial air-conditioning and heating units usually cut off their fan motors on the "off" cycle of the equipment. However, they are equipped with a fan switch that allows continuous operation.

Various other methods are used to control evaporator fan motors, but none is as common as the constant air flow and the fan cycling with equipment operation.

Condenser Fan Motor Controls

The condenser fan motor used on most commercial and industrial air-conditioning units is usually designed to maintain a constant head pressure by operating a certain number of condenser fans when the outside temperature drops below 75°F. This is accomplished by using thermostats to operate the condenser fan motors at certain outdoor temperatures or by using high-pressure switches to cycle the fans only when they are needed. The condensing temperature of a condensing unit should be maintained at or around 90°F for proper operation of the air-conditioning equipment.

Water-Cooled Condenser Interlocks

Water-cooled condensing units require an interlock in the condensing unit control circuit. The interlock ensures that the air-conditioning unit does not start unless the cooling tower pump and the cooling tower fan motor are operating, as shown in Figure 13.7. If the air-conditioning equipment were to operate without the pump operating, there could be damage to the compressor because of high head pressure.

Control System for Commercial Condensing Unit

A schematic diagram of a large condensing unit is discussed in this section. The unit is energized by applying 230 volts to the condensing unit. Unlike the small residential units, the control circuit of this unit is completely line voltage. The unit operates on 208 volts-3 phase-60 cycles.

The schematic diagram for the unit discussed in this section is shown in Figure 13.8. The unit is started by the energization of the control relay. This relay could be energized by a thermostat, pressure switch, or any other controlling device. The diagram will be discussed in four sections: the timer and timer relay circuits, the compressor circuits, the condenser fan motor circuits, and the safety controls of the circuit.

Timer and Timer Relay Circuits

Figure 13.9 shows the timer and timer relay circuits with the timer enclosed by a dashed line. This timer and timer relay arrangement is used by several manufacturers to protect the system from **short cycling** (rapid starting and stopping). The timer allows for a 5-minute delay from the time the unit shuts down to when it starts again.

The timer allows for a 5-minute delay from one cycle of the compressor to another cycle. The timer makes one complete revolution in a 5-minute period. The timer rotates 4 minutes and 45 seconds with the switches in the position from B to B1 and from A to A1. The timer runs because of the normally closed contacts of the timer relay TR(1) located in circuit 1. After the timer motor has run 4 minutes and 45 seconds, it changes the position of

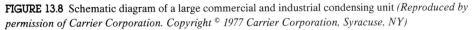

FIGURE 13.8 Schematic diagram of a large commercial and industrial condensing unit *(Reproduced by permission of Carrier Corporation. Copyright © 1977 Carrier Corporation, Syracuse, NY)*

FIGURE 13.9 Timer and timer relay circuits

the timer contacts from A to A2 and from B to B2 and stops the timer motor. This has no effect on the system until the control relay contacts in circuit 3 are closed.

When the control relay CR is closed, the timer relay becomes energized by circuit 3 through A2 to A of the timer. This action in turn closes the timer relay contacts TR(2) and TR(3) and opens contact TR(1). Once the timer relay is energized, the timer motor runs an additional 15 seconds and changes positions from A to A1 and from B to B1, which energizes the contactor C1 through the circuits of the timer. The timer motor does not run in this position because it is not being supplied with power. The contactor remains energized until the control relay CR is deenergized. The timer then restarts its cycle. Refer again to Figure 13.8 for control circuit details.

This timing method is one of several used as a protective circuit by the industry to prevent a heavy load from short cycling. This particular timer and timer relay circuit is one of the few methods that will reset itself automatically.

Compressor Circuits

The compressor circuit is shown in Figure 13.10 along with timer and timer relay circuits. The equipment is using a part winding motor to operate the compressor. A part winding motor has two separate motors, which are energized at different times (although close together), built in one housing. The

FIGURE 13.10 Compressor circuits and timer and timer relay circuits

second motor or winding is energized a few seconds after the first motor or winding by a time-delay relay. The motor is supplied by voltage through two contactors, C1 for the first motor or winding and C2 for the second.

The compressor circuits are equipped with circuit breakers CB1, CB2, and CB3 to protect the wiring and the motor. The power wiring is then split between the two contactors C1 and C2 that control the compressor. When the contacts of the contactor are closed, the compressor will operate.

The control circuit of the contactor is through the 208 volt-1 phase-60 cycle section of the diagram. This includes the CR contacts, the timer and timer relay circuits, the contactor coils, the time-delay relay, and the crankcase heater. The crankcase heater operates when the normally closed contacts TR(1) are closed and the timer relay coil is deenergized.

The compressor contactors control the operation of the compressor through the timer and timer relay. Assume the timer motor has run its 4 minutes and 45 seconds on the "off" cycle and the timer is now in the positions A to A2 and B to B2. When the control relay contacts close, the timer starts again and the timer relay energizes, closing contacts TR(2) and TR(3) and keeping the timer relay energized until the control relay contacts open. After the timer has run for the 15 seconds, it switches to positions A to A1 and B to B1. This action energizes contactor coil C1 and the time-delay relay (TDR). The time-delay relay closes about 2 seconds after C1 has energized, and this energizes C2. Now the compressor is operating.

Condenser Fan Motor Circuits

The condenser fan motors OFM1, OFM2, and OFM3 are shown in the circuits of Figure 13.11. The condenser fan motors are all controlled by the separate contactors OFC1, OFC2 and OFC3. The first condenser fan motor contactor OFC1 is energized by the closing of the control relay contacts in circuit 3. When the coil of contactor OFC1 is energized, the condenser fan motor OFM1 is started. The additional two condenser fan motors operate by the temperature surrounding the unit. If the temperature is high enough to close temperature switch TS1, then contactor OFC2 is energized and starts fan motor OFM2. Fan motor OFM3 follows the same operation as fan motor OFM2.

Circuit Safety Controls

The schematic diagram of the condensing unit in Figure 13.11 shows all the safety devices. The safety devices discussed here will not include fuses and circuit breakers.

All the safety devices are added in one circuit except the heater of the oil pressure switch, which is connected between A1 and L_2. If any of the safety devices open, the compressor is stopped by the opening of the TR(3) contacts when the timer relay is deenergized because of the broken circuit. When this occurs the timer must complete its "off" cycle rotation of 4 minutes and 45 seconds.

The high pressurestat (HP) and low pressurestat (LP) maintain the pressure in the refrigeration system within safe limits and open if the pressure exceeds the safe limit. The discharge line thermostat (DLT) opens to protect the compressor if the temperature of the discharge line exceeds a set temperature. The thermotector (TT) is an internal motor-winding thermostat that protects the compressor motor from overheating by opening if the motor temperature exceeds a safe level.

Legend

TB:	Term Block
TM:	Timer Motor
FU:	Fuse
TR:	Timer Relay
C:	Contactor
CB:	Circuit Breaker
OFC:	Outdoor Fan Contactor
OFM:	Outdoor Fan Motor
CH:	Crankcase Heater
OPS:	Oil Pressure Switch
TDR:	Time-Delay Relay
LLP:	Liquid Line Pressure Stat (field supplied)
DLT:	Discharge Line Thermostat
TS:	Temperature Switch
CR:	Control Relay
TT:	Thermotector
HP:	High Pressurestat
LP:	Low Pressurestat

FIGURE 13.11 Condenser motor circuits, compressor circuits, and timer and timer relay circuits

The oil pressure switch (OPS) opens if the oil pressure of the compressor is too low for safe operation. The oil pressure switch that makes up the oil safety switch is controlled by a heater and a pressure switch shown in a circuit from A1 to L_2. If the oil pressure switch remains closed, the heater heat up an element and open the switch of the OPS thermostat. If the oil pressure is high enough, the oil pressure switch (OPS) opens and cuts the heater out, allowing the unit to continue.

The schematic diagram covered in this chapter is not the most difficult diagram service personnel may encounter. However, it is a good example of what the service mechanic will come in contact with in the industry.

13.4 METHODS OF CONTROL ON ADVANCED SYSTEMS

There are many methods of controlling commercial and industrial systems. The total control system is often totally different from the control system of each piece of equipment, with the exception of the initiation and termination of the operation of the equipment.

The control initiation is different from the low-voltage control system because of the size of the contactors and the additional components. These systems sometimes use a low-voltage control system, but it would only be used to energize the main control system.

Most large commercial and industrial systems have a main control system that maintains the control of the total structure. It only initiates and terminates the operation of the equipment to deliver the heating and cooling medium at a desired temperature. The total control system takes care of the remaining conditions. Therefore, the air-conditioning control is accomplished by starting or stopping the equipment. The control system of the equipment takes care of the operation of the equipment.

Most systems are equipped with some method of capacity control, which allows the equipment to operate at from 100% to 33% of its capacity. The smaller commercial and industrial systems usually operate with a standard low-voltage staging thermostat that is capable of capacity control to some extent.

In this section we will discuss some basic control methods used on the larger commercial and industrial systems.

Small Commercial and Industrial Systems

The small commercial and industrial systems often use a simple low-voltage control system to maintain the desired temperature in the structure, with some staging accomplished by the thermostat. However, in many cases low voltage cannot be produced in sufficient strength to adequately close all the contactors or relays in a system. In this case, the low voltage would control a relay to supply 110 volts or 220 volts to the cooling equipment.

Figure 13.12 shows a common control system of a 5-ton split air-conditioning system with a fan coil unit. A control relay starts and stops the operation of the condensing unit by supplying line voltage to the unit. This control relay is operated by a thermostat. The thermostat also operates the evaporator fan motor by energizing a magnetic starter. After the initiation of the control of the entire system takes place, the condensing unit control system takes over and supervises the operation of the equipment. This control system has all the necessary safety devices for the major load and contains a timer with a relay to prevent rapid short cycling. The timer unit allows for a 5-minute delay between the cycles of the condensing unit.

Legend

C: Contactor
CH: Crankcase Heater
COMP: Compressor
CR: Control Relay
HP: High-Pressure Switch
HR: Holding Relay
IFR: Indoor Fan Relay
IT: Internal Thermostat
LP: Low-Pressure Switch
OFM: Outdoor Fan Motor
OFMC: Outdoor Fan Motor Capacitor
OL: Overload Relay
RC: Run Capacitor
SC: Start Capacitor
SR: Start Relay
TM: Timer Motor
TR: Transformer
———— Factory Wiring
- - - - - - Field Wiring

FIGURE 13.12 Control system of a 5-ton split air-conditioning system with a fan coil unit *(Reproduced by permission of Carrier Corporation. Copyright © 1977 Carrier Corporation, Syracuse, NY)*

It should be noted that many control systems will operate completely on low voltage and hence be very similar to the simple residential control systems.

Control Systems for Air-Cooled and Water-Cooled Packaged Units

Many air-cooled and water-cooled packaged units are designed strictly for commercial and industrial use. These packaged units are usually used in a free-standing installation put directly in the area that is to be conditioned, without any duct work.

This type of unit is equipped with a line voltage control system that is totally housed in the equipment.

The unit has a manual switch that allows it to be operated by this switch and a common line voltage thermostat. The three-position switch has an "off" position, a "fan only" position, and a "cool" position. The "cool" position of the switch is connected in series with a line voltage thermostat. It cuts the compressor on and off as needed by energizing a contactor or magnetic starter. If the unit is large, the evaporator fan motor is also controlled by a magnetic starter. The smaller evaporator fan motors, being single phase, are controlled by the switch.

All the safety devices are connected in series with the coil of the contactor. In most cases, crankcase heaters are used on this type of system and are energized on the "off" cycle of the compressor. Figure 13.13 shows a pictorial diagram of a common control system used on this type of equipment.

Pump-Down Control Systems

A large commercial and industrial air-conditioning condensing unit can be controlled by various types of initiation. The most common type of control is a pump-down system that incorporates a solenoid valve in the liquid line of the equipment. The valve opens and closes on the direction of the overall control system of the structure. The action of the solenoid valve determines if the condensing unit is to operate. If the solenoid valve is open, the unit will operate; if the solenoid closes, the unit will continue to operate until the low-pressure switch cuts the compressor off. Figure 13.14 shows a simple pump-down hookup of a condensing unit. The figure shows the solenoid valve being controlled by the thermostat and the action of the condensing unit controlled by the low-pressure switch. Some condensing units use a solenoid relay mounted in the unit to control the solenoid valves.

The pump-down control method is in common use on commercial and industrial systems because it will not allow refrigerant to feed to the coil on

FIGURE 13.13 Pictorial diagram of the control system of an air-cooled packaged unit with a remote condenser *(Reproduced by permission of Carrier Corporation. Copyright © 1977 Carrier Corporation, Syracuse, NY)*

Legend

T: Thermostat
LLS: Liquid Line Solenoid
LP: Low-Pressure Switch

FIGURE 13.14 Simple control system for the pump-down cycle of a condensing unit

the "off" cycle and migrate back to the compressor. The solenoid valve in this type of system can be controlled by almost any type of temperature sensing device that can be mounted in the desired portion of the system. Or the valve can be controlled by the equipment to prevent the valve from operating the equipment in unsafe conditions.

13.5 TOTAL COMMERCIAL AND INDUSTRIAL CONTROL SYSTEMS

The commercial and industrial control system is complex and difficult to understand because it controls virtually the total structure. It has the responsibility of controlling the zones of the structure, the air-conditioning equipment, the heating equipment, the evaporator fan motor, and the temperature of the entire structure. These control systems are usually designed and installed by one of the major control manufacturers.

The controls of a large system can be electric, pneumatic, or electronic. The electric control system is not used as often as the pneumatic and electronic because the devices must be larger and do not maintain as accurate control.

The pneumatic control system is operated by air pressure, which opens and closes switches and valves by air pressure. An example of a pneumatic control system is shown in Figure 13.15.

The electronic control system is shown in Figure 13.16. It controls the temperature in the zones by electronic devices.

These diagrams were shown here just as examples of some of the control

Legend

Ret A:	Return Air	T-2:	Room Thermostat
OA:	Outdoor Air	T1-1:	Discharge Air Thermometer
C-1:	Minimum Position Controller	V-1:	Hot Water Valve
DO-1	Outdoor Air Damper Controller	V-2:	Room Hot Water Valve
DO-2:	Return Air Damper Controller		(Convector)
D-1:	Outdoor Air Damper	TT-1:	Discharge Air Thermostat
D-2:	Return Air Damper	SV-1:	Air Solenoid Valve (Outside
PE-1:	Cooling Controller		Air)
T-3:	Mixed Air Thermostat	TEL-1:	Freeze Protector
RSCM-1:	Receiver Controller	-------	Air Lines Field Installed
T-1:	Return Air Thermostat	————	Supply Air Lines

FIGURE 13.15 Pneumatic control system *(Courtesy of Johnson Controls, Inc.)*

systems in use today. Their circuitry will not be discussed.

13.6 TROUBLESHOOTING CONTROL SYSTEMS

The service mechanic must develop a systematic method for finding the faults in a control system or in any electric part of the equipment. The responsibility of finding and correcting the problem in a control system falls on the person attempting to repair the equipment. This person is usually the service mechanic. It is no easy task to diagnose problems in electric

Legend

DOH:	Outdoor Air Damper	TR:	Transformer
TEL:	Outdoor Air Thermostat	TEC:	Main Controller
LL:	Low-Limit Thermostat	N:	Outdoor Air-Adjustment Relay
VB:	Hot-Water Valve	RN:	Staging Relay
VOH:	Valve Operator	TES1:	Room Sensor
SV:	Cooling Solenoid Valve	TES2:	Discharge Air Sensor

FIGURE 13.16 Electronic control system *(Courtesy of Johnson Controls, Inc.)*

systems, especially the larger commercial and industrial equipment. The smaller residential and commercial equipment, fortunately, is built and designed with a simpler control system and fewer components than the larger equipment.

Service personnel must understand how, when, and why a certain piece of equipment operates as it does. The service mechanic has many important tools that can be used to find and correct the problem in the equipment. The wiring diagram of the equipment is extremely important to the service mechanic because it tells how the equipment operates. The diagram also tells the service mechanic the components in the electric system and their location in the system. The schematic diagram is usually the easiest tool to use for diagnosing problems. However, in many cases manufacturers will give a pictorial diagram or a combination of a pictorial and schematic, with the equipment. No matter what type of diagram is furnished with the equipment, the service mechanic must find and repair the malfunction in the equipment. The voltmeter, ammeter, and ohmmeter are all important tools in accomplishing this task.

It is essential that service personnel be able to read wiring diagrams and use electric meters to diagnose electrical problems and make the necessary repairs. Electric components are usually the cause of any electrical problems in the system. Therefore, the service mechanic must be efficient at correctly diagnosing the condition of the electric components of the system, such as the compressor, electric motor, relays, contactors, transformers, thermostats, and pressure switches. Of course, the mechanic must at all times be careful not to overlook the small things that homeowners or inexperienced service mechanics could have done to an electric system, such as changing wires, jumping controls out, and removing components they felt were unnecessary.

Diagnosis of Electric Components

The diagnosis of electric components is a fairly easy task for the mechanic who understands the particular device (see the earlier chapters in this book). In many cases, service mechanics assume the jobs of parts changers, instead of actually diagnosing each component that has a possibility of being faulty. A good service mechanic will change some electric components that are perfectly good, but this should happen only rarely.

Electrical problems are very common throughout the industry and are said to cause 80% to 90% of the reported system failures. Thus, it is important for service personnel to be able to diagnose and correct electrical problems in a system.

Motors. The electric motor is the largest load used in most heating, cooling, and refrigeration systems. The compressor motor is the most expensive component of the system and care should be taken to correctly diagnose this component because of the expense that can be needlessly incurred. In fact, all electric motors are expensive and should be accurately diagnosed. For details about electric motors, refer again to Chapters 7 and 8.

Switches. Electric switching devices can cause many of the electrical problems encountered by service mechanics. These devices are used to control some load in the system. The relay and contactor are basically the same device except for their size and ampacity. These devices must be checked for two possible malfunctions. First, the contacts or mechanical linkage can be bad. Second, the coil that closes the contacts can be faulty. These devices are easily diagnosed for problems if the mechanic understands their operation and construction.

Thermostats and pressure switches are switches that are controlled by the temperature of the medium surrounding them or the pressure at the point they are connected to the system. A simple thermostat or pressure switch can easily be diagnosed because they have only one set of contacts. But the mechanic must be absolutely certain that the problem is not due to some other device that is malfunctioning.

The low-voltage thermostat used on the smaller systems is more difficult to diagnose because of its complexity and the fact that it is used for several purposes on a heating and cooling system. The service mechanic should be careful not to condemn as faulty a thermostat that is merely out of calibration by a few degrees.

Many other electric components are used in control systems that have not been covered in this review. However, they should be easily diagnosed when the mechanic studies the component along with the wiring diagram.

Diagnosis of Control Systems

In many cases, the components are at fault when the system is not operating properly. But it is sometimes difficult to identify the faulty component unless the mechanic uses the wiring diagram to identify the circuit that is not operating properly and then goes to the load or to the device that controls the load. A mechanic can usually identify the load that is not operating properly. Therefore, the circuit should be checked for any malfunctions or faulty components.

Troubleshooting the electrical control circuits in an air-conditioning, heating, or refrigeration system is basically common sense. It is essential that the service mechanic understand how the unit functions electrically.

With a good knowledge of how the unit functions electrically and a good troubleshooting approach, the service mechanic can eliminate guesswork and parts changing when servicing a unit. Observing the unit in operation will lead the mechanic to the electrical circuits that can be eliminated as trouble sources. This process of elimination is based on a systematic analysis of the unit's operation. Most service mechanics quickly perform this task mentally once the operation of the unit has been observed. Troubleshooting the electrical circuitry of a unit is nothing more than systematically eliminating the circuits that are operating correctly. For example, if an air-conditioning system has three major loads, as shown in Figure 13.17, and two of the three loads are operating correctly, then the mechanic only has to check the circuit that is controlling the load along with the load that is not operating properly. In Figure 13.17, if the indoor fan motor is not oper-

Legend

C:	Contactor
COMP:	Compressor
CRC:	Compressor Running Capacitor
CFM:	Condenser Fan Motor
CFMC:	Condenser Fan Motor Capacitor
IFR:	Indoor Fan Relay
IFM:	Indoor Fan Motor
HP:	High-Pressure Switch
LP:	Low-Pressure Switch
TR:	Transformer

FIGURE 13.17 Total control circuit in a small residential air-conditioning system

ating, then the mechanic should concentrate on the indoor fan motor and its controlling circuits. The service mechanic must be able to recognize and separate the load circuits.

Residential Systems.   Figure 13.18 shows a basic circuit in a residential air conditioner. The circuit contains the contactor and the compressor, with the contactor controlling the load. If the compressor is not operating, it is logical to assume that the problem is with the power supply to the contactor or the unit. The first check should be the incoming power to the contactor, usually to a terminal board. If it is available, continue the voltage check to the compressor. If voltage is not available to the compressor, check the contacts and make certain of their position. If the contactor is closed and the compressor is being supplied with the correct power, diagnose the compressor, making certain to consider the overload if it is internal. Check the contacts of the contactor if the contactor is closed but no voltage is going through its contacts. However, if the contacts are not closed, you will have to look at the circuit that controls the contactor coil.

Total Circuit.   The total control circuit of a residential air-conditioning system is shown in Figure 13.17, including the total wiring diagram. The methods used in the previous paragraphs can be followed here. The low-voltage control circuit can be checked out if the contactor is not closing, assuming the compressor is good and the contactor is not closed. The first check that should be made is the low-voltage power supply to ensure that voltage is available to the control circuit. The line voltage power supply has been checked and is correct.

Once it is determined that voltage is available to the control circuit, the only items left to check are the devices in the circuit. In the control circuit there are four devices that should be checked: the low-pressure and the high-pressure switch, the thermostat, and the contactor coil. The contactor coil can be checked by reading the voltage available to it. If 24 volts are available but the contactor is not closed, then the coil or mechanical linkage of the contactor is bad and should be replaced. The other three components should be checked for an open circuit. This can be done easily with an ohmmeter or voltmeter. Do not overlook any problem that could have the high-pressure or low-pressure switches opened (low refrigerant charge, dirty condenser, or a bad condenser fan motor).

Hopscotching.   One frequently used procedure for troubleshooting an electric circuit is called **hopscotching**. Figure 13.19 shows an electric circuit that we will use as an example to follow the procedure. The contactor coil in the circuit closes, starting a compressor. Let us assume that we are getting volt-

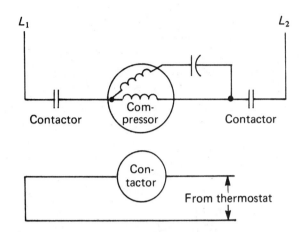

FIGURE 13.18 Basic circuit in a residential air-conditioning system

FIGURE 13.19 Simple electric circuit for hopscotching

age to the system but not to the contactor coil. If the correct voltage is not available to the coil, one of the switches in the contactor coil circuit would have to be open. The question is which one. First connect one lead of a voltmeter to L_2. The other lead of the voltmeter should be placed at point A in Figure 13.19. If line voltage is read, the switch at A is closed and not at fault. Then hop to points B, C, and D with the meter leads. At the point where voltage is not read, that switch is open.

This procedure of troubleshooting is only one of many used in the industry. Good service technicians will usually develop their own procedures as they gain experience with the various types of equipment.

Industrial Systems. Commercial and industrial control systems have many more components and are more complex than the simple residential control systems. Therefore, it is more difficult to diagnose and correct the problems in these circuits. First check the power supply to the equipment and the power supply to the control circuit. If there is power, use the wiring diagram and follow a systematic method for troubleshooting the system. The complexity and number of components should not cause you any concern. Each circuit of the diagram can be and should be treated individually.

Service personnel soon learn their own methods of troubleshooting equipment that allows for faster and more efficient service calls. The one item that takes its toll on service mechanics is a callback. A callback is a return trip to a piece of equipment that a mechanic has not fixed properly. This hurts the efficiency of the mechanic as well as the employer because of the added expense, which cannot be charged to the consumer. Service mechanics should try to eliminate all possibility of callbacks. If a mechanic does the right kind of job on the first call, there is no callback.

One procedure for troubleshooting any air-conditioning, heating, or refrigeration system is the following:

1. Check the power supply to the equipment.
2. If no power is available, correct the problem.
3. Make a thorough check of the system to determine which load or loads are not working properly.
4. Locate the loads on the wiring diagram that are not operating properly and begin checking these circuits first.
5. Each device in that circuit is a possible cause of the problem and should be checked until the faulty device is found. (Do not disregard the load in the circuit.)
6. It may be necessary to move to other circuits to check the coils of some of the devices.
7. Never overlook the internal overload devices.

Most manufacturers make a troubleshooting chart available to air-conditioning dealers that handle their line of equipment. Figure 13.20 on pages 317-318 shows a troubleshooting chart of one manufacturer for a total cooling system. Another type of troubleshooting chart is shown in Figure 13.21 on pages 319-320 for the heating and cooling cycle of a heat pump. Other manufacturers develop troubleshooting charts on each different model of equipment they produce. If these charts are available, it is advisable that they be used, especially if you are just beginning your service career.

The heating, cooling, and refrigeration industry depends as much on the service aspects as on the selling of new equipment. Service work is a necessity if a company is going to be successful. No consumer wants an air-conditioning and heating system if he or she cannot get the proper service and maintenance. Remember, too, that service personnel always reflect the

TROUBLE-TRACER*
Summer Air Conditioning and Heat Pump

Satisfactory operation of an adequately sized air conditioning system depends on proper discharge pressure, suction pressure, current draw and air distribution. The service man must be able to determine these in order to properly diagnose system operation

Center of chart states basic trouble. Begin there.

Adjacent partial circles state specific troubles that could cause basic trouble. Determine which of these specific troubles exist before proceeding further.

Remaining partial circles state various contributing troubles to specific trouble. Determine contributing trouble or troubles before attempting correction.

Numbers in chart are for explanation and guidance. Where no numbers are shown, the statement is considered to be self-explantory.

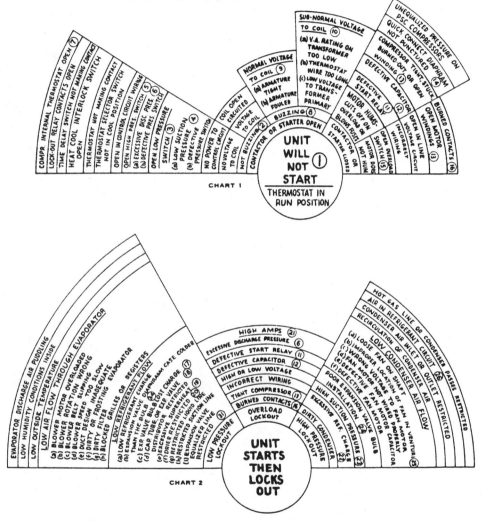

FIGURE 13.20 Manufacturer's troubleshooting chart *(Courtesy of Tappan Air Conditioning Division)*

employer when they are sent out on the job. They should always try to present themselves in a manner that will sell them as well as their employer. The appearance and attitude of a service mechanic are just as important as the ability to diagnose and correct the problem that the homeowner is having.

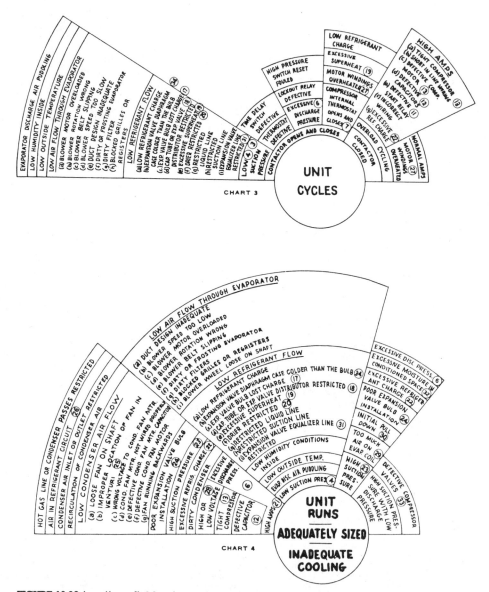

FIGURE 13.20 (continued) Manufacturer's troubleshooting chart *(Courtesy of Tappan Air Conditioning Division)*

FIGURE 13.21 Troubleshooting chart for heating and cooling cycle of heat pump *(Reproduced by permission of Carrier Corporation. Copyright © 1977 Carrier Corporation, Syracuse, NY)*

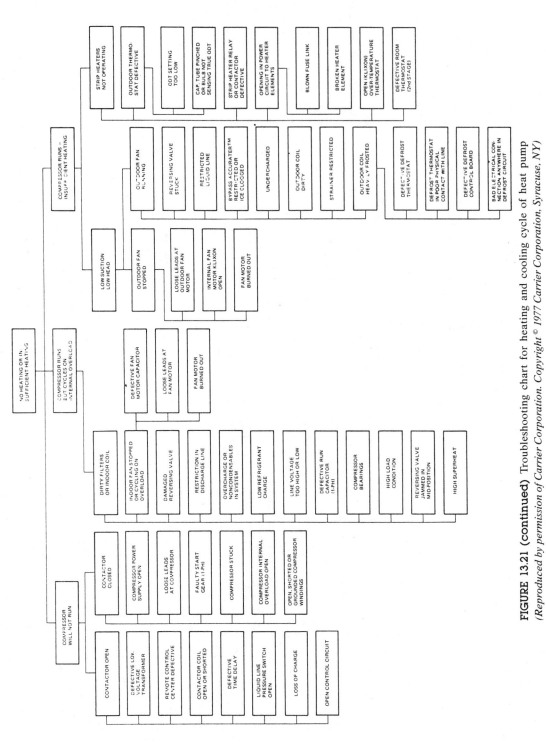

FIGURE 13.21 (continued) Troubleshooting chart for heating and cooling cycle of heat pump (*Reproduced by permission of Carrier Corporation. Copyright © 1977 Carrier Corporation, Syracuse, NY*)

SUMMARY

There are certain similarities among control systems used on the smaller heating and air-conditioning equipment. The contactor in most cases will control the compressor and the condenser fan motor. In some cases, the condenser fan motors are controlled by a thermostat to maintain a constant head pressure of the system, but usually a contactor controls a two-speed fan motor. The evaporator fan motors are usually controlled by some temperature on the heating cycle and an indoor fan relay on the cooling cycle. The safety components are in series with the contactor controlling the compressor because it is the largest, most important, and most expensive load in the heating and air-conditioning system. From a familiarity with the common basic control circuits, an installation or service mechanic can easily adapt to the different control systems used in the industry. Usually some common type of control is used in all control circuits.

The advanced control system required on the larger commercial and industrial equipment is more complex and sophisticated than the smaller equipment controls because of its size and operation. In many large structures, the control system controls the structure and only maintains a certain temperature for the heating and cooling medium. Therefore, the main control system signals for the cooling or heating equipment to operate. Once the equipment has been started, the control system of the equipment supervises the operation of the unit, including the capacity control.

There are basically two methods used to start condensing units and air-conditioning equipment. The pump-down method uses a solenoid valve to start and stop the flow of refrigerant to the evaporator. The other method used to start the equipment is by supplying a certain voltage to the equipment when it needs to be started. Large control systems have many basic circuits that are common to equipment throughout the industry.

The servicing of any air-conditioning or heating system is important because sooner or later the equipment will malfunction and need attention. Service mechanics must develop a systematic approach to diagnosing the faulty equipment and making the necessary repairs. A good beginning for any service mechanic is to be able to understand and diagnose all the components of the system. Service mechanics should use the many tools that are available to them when servicing equipment, such as wiring diagrams, testing equipment, and any service manuals that are available. No matter how large or how small an air-conditioning, heating, or refrigeration system is, the mechanic can break down the control system and the electric system into basic circuits and then check the circuits on an individual basis. This is much easier than attempting to treat the total diagram as one large circuit. By using a systematic approach to troubleshooting electric systems, service mechanics can accomplish any troubleshooting task within reason.

QUESTIONS

1. Explain briefly the operation of the unit in Figure 13.22.

2. Explain briefly the operation of the unit in Figure 13.23.

3. Outline one procedure for troubleshooting a control system.

4. Name three components that would be common to all control systems.

5. True or false. All evaporator fan motor circuits are alike for the smaller residential systems.

6. On residential systems the condenser fan motor is controlled by _____.

7. True or false. Safety controls used in the smaller systems are connected in parallel with the contactor coil.

Legend

C:	Compressor Contactor
COMPR:	Compressor
F:	Fan Contactor
FM:	Fan Motor
HPS:	High-Pressure Switch
IT:	Internal Thermostat
LPS:	Low-Pressure Switch
OL:	Overload Protector
SW:	Selector Switch
T:	Thermostat
⬡ :	Component Connection, Marked
◇ :	Component Connection, Unmarked

▢ :	Terminal Board Connection
———	Control Wires
▬▬▬	Power (factory) Wires
- - - -	Field Wires
— — —	Alternate CSR Wiring

Notes

1. Fan motors furnished with inherent thermal protection; compressor motors thermal and overload protection.

2. Control terminal block for connection of Class 1 control circuit.

FIGURE 13.22 Diagram for Questions 1 and 21 *(Reproduced by permission of Carrier Corporation. Copyright © 1977 Carrier Corporation, Syracuse, NY)*

Legend

C: Contact
CH: Crankcase Heater
COMP: Compressor
CR: Control Relay
DFR: Defrost Relay
DFT: Defrost Thermostat
DT: Defrost Timer
IFM: Indoor Fan Motor
IFMC: Indoor Fan Motor Capacitor
IFR: Indoor Fan Relay
OFM: Outdoor Fan Motor
OFMC: Outdoor Fan Motor Capacitor
QT: Quad Terminal
RC: Run Capacitor
RVR: Reversing Valve Solenoid
TB: Terminal Block
TRAN: Transformer
OTT: Over Temperature Thermostat
LLPS: Liquid Line Pressure Switch
ST: Start Thermister

FIGURE 13.23 Diagram for Questions 2 and 22 *(Reproduced by permission of Carrier Corporation. Copyright © 1977 Carrier Corporation, Syracuse, NY)*

8. Why is capacity control used on the larger systems?

9. The part winding motor is used to _____.

10. True or false. In the larger systems, fan motors are generally designed to operate at all times.

11. A constant head pressure is maintained in a system by the _____.

12. In the larger systems, a low-voltage control system would be used to _____.

13. What is the purpose of the main control system in the larger industrial and commercial units?

14. What is a pump-down control system? How does it operate?

15. The controls of a large system can be _____, _____, or _____.

16. True or false. The compressor is the most expensive component of an air-conditioning system.

17. What is the first check that should be made in diagnosing a basic control system?

18. Describe what is meant by the term "hopscotching."

19. Outline one procedure you might use for troubleshooting an air-conditioning, heating, or refrigeration system.

20. What part does the process of elimination play when troubleshooting an air-conditioning unit?

21. Use Figure 13.22 to answer the following questions:
    a. What is the problem with the unit if the contactor is closed but the compressor is not operating?
    b. What safety devices could prevent the contactor from closing if the thermostat, selector switch, and the two overloads are closed?

22. Use Figure 13.23 to answer the following questions:
    a. What type of thermostat is used in the diagram?
    b. What are the possible problems with the control system if the control relay is energized but the compressor is not operating?
    c. What is the problem with the unit if it will not come out of the defrost cycle?

LAB SECTION REFERENCE

For experiments and activities dealing with material covered in this chapter, refer to Lab Number 15, Troubleshooting Air-Conditioning Systems.

Solid-State Controls and Systems

INTRODUCTION

The refrigeration, heating, and air-conditioning industry has moved rapidly to incorporate solid-state control circuits into many of the control systems being used presently. Vast improvements made in solid-state components have revolutionized the industry, yielding smaller, more accurate, and more diversified control systems. Solid-state control systems allow better control for the entire heating and air-conditioning system. Along with better control, many other advances have been made: in motor speed control, in motor protection, and in defrosting systems for heat pumps and commercial refrigeration.

Major improvements have been made in large rooftop multizone units by using solid-state control systems. Controls for these units in the past were large and bulky and often did not give adequate control to each zone of the building. Electronic controls have almost eliminated erratic control of individual zones in this type of application. Many of these units also incorporate a method for controlling the fan speed, depending on the load of the building, by using solid-state speed control.

Commercial refrigeration control systems have also been affected by the influx of solid-state control systems. Many modern commercial refrigeration systems are controlled by solid-state systems that give better control to the entire system. Defrosting systems for this type of equipment have made the largest step forward by using solid-state controls.

Smaller refrigeration, heating, and air-conditioning systems have also been affected by the wide use of solid-state controls. Most small systems use solid-state components for a specific function, but a few use a total solid-state control system. Many manufacturers use some type of solid-state control for the protection of the compressor motor used in air-conditioning and refrigeration systems. Solid-state defrost controls are included on many heat pumps. The demand for energy efficiency has moved the industry toward better control of furnaces, which is accomplished in the newer systems by solid-state controls. Several new heat pumps that use total solid-state control systems have been introduced to the industry. These systems provide better control and efficiency for the consumer.

These new solid-state control systems are not unlike conventional controls in that they require servicing. Although the term "solid state" may scare many service people presently in the field, there is no need to be intimidated by the new electronic controls. The new solid-state control systems are nothing more than a group of circuits used to control a system. Furthermore, there is a similarity between these systems and the old conventional systems.

The heating, air-conditioning, and refrigeration technician must become proficient at troubleshooting solid-state modules now appearing in the field. The main thrust of the mechanic's troubleshooting will be aimed at the module rather than each individual solid-state component housed in the module. In most cases, the mechanic will determine that a solid-state module is faulty and will change the entire module rather than attempt to troubleshoot the module components. Most heating, air-conditioning, and refrigeration companies are not equipped to repair the module. Therefore, once the determination is made that the module is bad, it is merely replaced with no further testing. With the influx of solid-state modules into the industry, it is imperative that mechanics be able to troubleshoot them.

14.1 ELECTRONIC SYSTEM COMPONENTS

Electronic systems have seen vast improvements over the past thirty years. Along with being large and bulky, the old vacuum and gas-filled tubes were made of many delicate parts. The newer solid-state devices are smaller and more durable than the old vacuum tubes. Figure 14.1 shows a size comparison of a vacuum tube and a transistor. The size and durability of solid-state devices have made it possible to incorporate them into many heating, air-conditioning, and refrigeration control circuits.

Semiconductors

Many terms and definitions are used in electronic control circuits, but it is not necessary for the heating, air-conditioning, and refrigeration mechanic

FIGURE 14.1 Size comparison of a vacuum tube
and a transistor

to become an electronics technician to efficiently service and repair this
new breed of control. However, the mechanic should have a basic understanding of major electronic components.

Although the terms "solid state" and "semiconductor" may refer to the
same device, they indicate different properties of the device. "Solid state"
refers to a physical description of the component. "Semiconductor" refers
to the electrical conductivity of the materials used in solid-state components.

A detailed description of semiconductor operation is not necessary to
understand the function and operation of solid-state control modules. We
will present only a brief introduction here.

The conductivity of materials plays an important part in the action of
semiconductors. For example, good electrical conductors are copper, silver, and aluminum; insulators like plastic, glass, and mica are poor conductors. **Semiconductors**, as their name implies, are half conductors. Some of
the materials used as semiconductors have a resistance almost halfway
between that of glass (insulator) and copper (conductor).

Semiconductor devices are versatile. They can provide a wide range of
electrical characteristics, making it easier to apply solid-state components
to air-conditioning control applications. The one disadvantage of semiconductors is that they cannot stand surges in voltage. In many applications, it
is necessary to protect these devices against adverse conditions.

Solid-state devices in their simpler forms date back to the 1930s. In the
1940s transistors were developed by two scientists at the Bell Telephone
Laboratories. The transistor is a combination of two other semiconductor

devices: the thermistor and the varistor. The thermistor is a solid-state device whose resistance decreases with an increase in temperature. The varistor's resistance decreases with an increase in voltage.

The most common materials used to make semiconductors are germanium and silicon. At the present time, silicon is used more than germanium. The atomic structure of silicon and germanium is important to the semiconductor material because it basically controls their operation. So that semiconductor materials can be readily utilized, certain impurities must be added to the basic element to improve the atomic structure, allowing the semiconductor to do its job. The silicon and the germanium atoms are extremely pure, and any element added to them is an impurity. The most commonly used impurities are aluminum, arsenic, antimony, boron, and gallium.

When certain impurities are added to silicon and germanium, the atomic structure of the semiconductor changes. This change provides the semiconductor with free electrons that can easily be moved and allow current to flow. Semiconductors with this type of impurity added are called **N- (negative) type** materials. If a different impurity is added to the semiconductor material, the atomic structure will be altered to allow electrons into the semiconductor. These vacancies are referred to as **holes**. The holes allow electrons to easily pass through. This type of material is called a **P- (positive) type** semiconductor. When these two types are put together, they become a transistor, rectifier, or diode. Figure 14.2 shows some common examples of P- and N-type materials being combined to form components.

Diodes and Rectifiers

The diode and the rectifier are the simplest of the solid-state components. The only difference between the two is their size; they are basically the same device. A component that is rated at less than 1 ampere is called a **diode**. A similar component rated above 1 ampere is called a **rectifier**. The common diode and rectifier conduct electricity in one direction much better than they do in the other direction. This is a basic characteristic for rectifying alternating current to direct current.

Diodes and rectifiers are composed of P-type materials bonded to N-type materials. They are made by taking a solid-state material that has free electrons and bonding it to another material that has a shortage of electrons.

When the two types of material are bonded together, a solid-state component is produced that will allow electrons to flow in one direction and act as an insulator when voltage is reversed. Electron flow of the device will be from the N-type to the P-type material. Figure 14.3 shows a basic drawing of a diode and rectifier. Figure 14.4 shows the symbol for a diode and rectifier.

Electron flow

FIGURE 14.2 P- and N-type materials combine to form components.

FIGURE 14.3 Drawing of a diode and rectifier

Figure 14.5 shows an electrical circuit using two rectifiers to change alternating current (AC) to direct current (DC) voltage. The rectifiers allow current to flow in one direction but will not allow the normal alternating current reversal, thus rectifying the voltage.

Transistors

Transistors are composed of N-type and P-type materials. Their exact com-

FIGURE 14.4 Symbol for a diode and rectifier

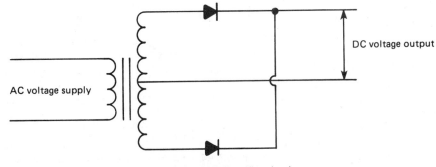

FIGURE 14.5 Rectifier circuit

position is dependent on the type of material used in the middle. If N-type material is sandwiched between two P-type materials, the transistor is referred to as a **PNP transistor**. If P-type material is sandwiched between two N-type materials, the transistor is referred to as an **NPN transistor**. Figure 14.6 shows the common symbols for the PNP and NPN transistors. The basic PNP or NPN transistor has three parts: the base, the emitter, and the collector. The transistor can be connected in the circuit in several different ways.

The transistor needed for a certain job would depend on the circuit, the application, the voltage, the current the device must handle, and the temperature. The transistor is used as a switch or to amplify a signal. In a radio, the transistor will take a weak signal and increase it until there is enough power so that an audible sound can be transmitted to a speaker. In air-conditioning control systems, a transistor will amplify a small signal obtained from a thermistor to provide enough power to do useful work, such as closing a relay or a contactor or driving a small motor.

A simple control circuit used for temperature control would have a thermistor in the input circuit, which varies its resistance with a change in temperature. A change in the resistance of the input circuit would cause a change in current to the transistor base, and the change would be amplified through the transistor. This amplified signal could have sufficient current to operate a switch or energize or deenergize a relay or contactor. Figure 14.7 shows a simple circuit for temperature control.

NPN transistor

PNP transistor

FIGURE 14.6 Symbols for an NPN and a PNP transistor

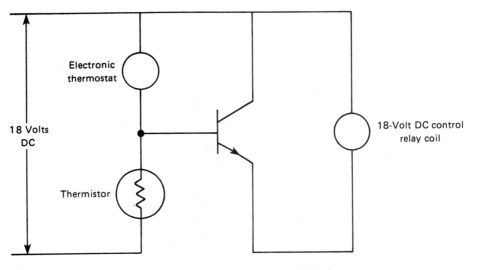

FIGURE 14.7 Temperature control circuit

14.2 BASIC ELECTRONIC CONTROLS

In the heating, air-conditioning, and refrigeration industry, many applications of solid-state controls are used as one-function devices. A **one-function solid-state device** is used in a control system primarily to control or protect one function in the operation of the equipment. For example, some of the one-function controls are defrost controls on a heat pump and time-delay devices to prevent a compressor from short cycling, to provide motor protection, and to act as starting relays on a split-phase hermetic compressor. Although many solid-state devices are used for numerous and different controls in heating, air-conditioning, and refrigeration control systems, in this section only a few will be covered.

Solid-State Timers

One of the most popular solid-state devices is the timer. Timers are available in many different forms and serve various functions in the electrical circuitry of air-conditioning, heating, and refrigeration equipment. Solid-state timers are used in the same manner as the conventional time-delay relay but with much more flexibility. Solid-state timers can delay on the closing or opening of an operating control. They are available with a set delay, used mainly by manufacturers on new equipment, and an adjustable delay, installed by field personnel on equipment for specific functions, such as delaying two heavy loads from starting at the same time or preventing short-cycling of equipment. Most timers are adjustable over a wide range of

FIGURE 14.8 Adjustable solid state timer with wiring diagram *(Copyright © 1984 Motors & Armatures, Inc. Used by permission.)*

timing cycles and are suitable for a variety of control circuit voltages. An adjustable solid-state timer is shown in Figure 14.8. These devices are compact and easy to install.

Anti-Short-Cycling Devices

One of the simplest solid-state devices in common use today is a control that prevents a compressor from starting more than once in a certain period of time. These devices can delay the starting time from 30 seconds to 5 minutes between cycles. An anti-short-cycling device is shown in Figure 14.9. This device is like many other simple solid-state devices; it should delay the starting of the specified component for the designated period of time or the device is not working properly. Figure 14.10 shows a control circuit with an anti-short-cycling device. The operation of these devices uses a solid-state material whose cold resistance is high and will not allow voltage to pass. But as the device heats, the resistance decreases, allowing voltage to pass through the device to energize a contactor or relay.

For Electric Motors

A solid-state device is now being used to assist the permanent split-capacitor motor in starting. In many cases, this device can be used instead of a potential relay and a starting capacitor. Figure 14.11 shows the start-assist kit.

FIGURE 14.9 Anti-short-cycling device

FIGURE 14.10 Control circuit with an anti-short-cycling device

FIGURE 14.11 Start-assist kit

This solid-state device uses a material that when wired in parallel with the running capacitor of a PSC motor (compressor), provides an increase in its starting torque. This material performs much like a small starting capacitor; it momentarily increases the current in the motor starting winding. As the material heats up, its resistance increases quickly to a point where it becomes a nonconductor and the motor returns to PSC operation. This device requires a 3-minute cool-down period between starts. To check this device, the service mechanic need only determine if it is dropping out of the circuit or if it is being energized on start-up. If the device is performing in any one of the above ways, it should be replaced.

The solid-state starting relay used on small split-phase motor compressors is also simple in its operation. The device is used to drop out the starting winding of a split-phase hermetic compressor motor. Two devices are shown in Figure 14.12; they are simple to install and service. Figure 14.13 shows a device installed on a fractional-horsepower hermetic compressor motor.

The solid-state starting relay is actuated on a thermal basis through the use of a semiconductor material. This material is of a given cold resistance. As the compressor starts, the material heats up quickly and becomes a nonconductive material, dropping the starting winding out of the circuit. The use of this device is limited to split-phase motors; it should not be used on capacitor-start motors and rapid-cycling compressors because a cool-down period is required for proper operation.

For troubleshooting the device, it is only necessary to determine if the device is dropping out the starting winding of the hermetic compressor motor. If the device is not energizing or dropping out the starting winding, it is faulty and should be replaced.

For Heat Pumps

Many heat pumps use a solid-state defrost control module that initiates and terminates the defrost cycle of a heat pump. This defrost control keeps the outdoor coil of a heat pump clear of frost, making the unit more efficient. A solid-state defrost module is shown in Figure 14.14. This defrost module uses two thermistors to initiate and terminate the defrost cycle and a pres-

FIGURE 14.12 Solid-state starting relays

FIGURE 14.13 Solid-state starting relay installed on a compressor

FIGURE 14.14 Solid-state defrost module

FIGURE 14.15 Wiring diagram of a solid-state defrost module

sure switch, as a safety, to terminate the cycle only if the module fails to terminate the defrost cycle.

One of the solid-state defrost modules in present use is very effective in preventing a buildup of frost on the outdoor coil of a heat pump. This module controls an 18-volt DC relay that stops the outdoor fan motor and energizes the reversing valve. When the reversing valve is energized, the refrigerant cycle runs in the cooling mode of operation and defrosts the outdoor coil.

The module is shown schematically with the DC relay and the two thermistors in Figure 14.15. The module has eight connections: two to the unit's 24-volt power supply; two to the coil of the 18-volt DC relay; two to the coil thermistor; and two to the thermistor in the airflow across the coil. The thermistors sense the temperature of the outdoor coil and the air temperature across the coil. To initiate the defrost cycle, the coil temperature must be below 30°F. At that time the module would feed 18 volts DC to the defrost relay, starting the defrost cycle by energizing the defrost relay and putting the unit in the cooling mode. The power is supplied to the module by the AC terminals from the 24-volt AC power supply of the unit.

Once the coil has become defrosted, the air temperature crossing the coil and the coil temperature would be high enough to bring the unit out of the defrost cycle, sending the system back to the heating mode of operation. Figure 14.16 shows the schematic of the heat pump.

The troubleshooting procedure for the defrost system is simple and easy to perform. The mechanic must first determine what the module should be doing and what it is actually doing. If the coil or air temperature is high enough, the system should not go to the defrost cycle. If the system defrosts at a higher temperature than it should, it is a faulty module or has faulty thermistors. To determine which is faulty, the mechanic should use the manufacturer's specifications. If the resistance of the thermistors is correct for the temperature indicated, then the defrost module is faulty and should be replaced if it is sending 18 volts DC to the defrost relay. If the module is not sending 18 volts DC to the relay and the unit is in the defrost mode, the defrost relay is stuck in the defrost position. If all conditions exist to start defrost and the module is not sending 18 volts DC to the defrost relay, then the module is faulty, assuming that the thermistors have checked out to correspond to the temperature. Troubleshooting solid-state devices is often a process of eliminating the good portions of the particular system and spotting the defective portion.

For Motor Protection

Motor protection is another area in which solid-state devices are used effectively. These types of safety control systems use thermistors to indicate the temperature of a hermetic compressor motor. The solid-state motor protection is fast acting and sustains periodic and moderate overloads without nuisance trip-outs.

Heat is the main enemy of an electric hermetic motor. The amount of heat that is produced and held in a hermetic motor is the result of load heat and motor efficiency. Some of the input to an electric motor is converted to heat as well as to useful work. Motor life is determined largely by the conditions that are imposed on the motor. Therefore, it is essential that the hermetic motor be effectively protected. The solid-state motor protector can effectively protect the motor within a reasonable period of time.

The most common overload protection used is a single-module, three-sensor system. Figure 14.17 shows the module and a single sensor used in the system. The sensors are inserted in the hermetic motor winding, making it more effective than conventional overload systems. The sensor is resistant to oil and refrigerant contained in a refrigeration system.

Figure 14.18 shows the sensor location in a three-phase hermetic motor.

DESCRIPTION		
KEY	COMPONENT	LOCATION
B1	compressor	2
B2	motor-outdoor fan	2
B3	motor-indoor blower	3
C1-C2-C3	capacitors	2.3
CH	heater crankcase	1
F	fuse	1
K1	contactor-compressor and outdoor fan (2P-NO)	2
K2	control-defrost	2
K3	relay-indoor blower (SPST)	3
K4	relay outdoor fan & defrost (3P-DT)	1-2-3
L1	valve-reversing	2
RT-C	thermistor-coil	2
RT-A	thermistor-air	2
S1	switch-disconnect (not provided)	–
S2	thermostat-room	2-3
S3	switch high pressure	2
S4	switch-loss of charge	2
S5	switch-defrost	3
T1	transformer-thermostat (208-230v pri.) 24v sec.	1
HVJB	high voltage junction box	–
LVJB	low voltage junction box	–
S6	thermostat-low ambient	2
G	equipment ground	–

COLOR CODE CHART	
BK	black
BL	blue
W	white
R	red
O	orange
Y	yellow
BR	brown

1. May not be used on early production units.

Thermostat Heat Anticipation

2. 1st stage—fixed
2nd stage—set according to electric heat

FIGURE 14.16 Schematic of a heat pump with solid-state defrost control *(Redrawn with permission of Lennox Industries, Inc.)*

FIGURE 14.17 Solid-state model, three-sensor system

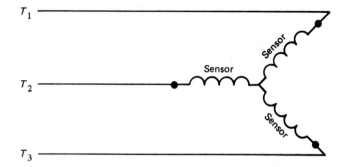

FIGURE 14.18 Sensor location in a three-phase motor

The sensors effectively sense the temperature of a hermetic motor. Figure 14.19 is a graph showing a sensor trip-out point and reset point. The resistances of the sensors increase as the temperature rises. This action causes the module to react and break a set of contacts to open the control circuit of the compressor.

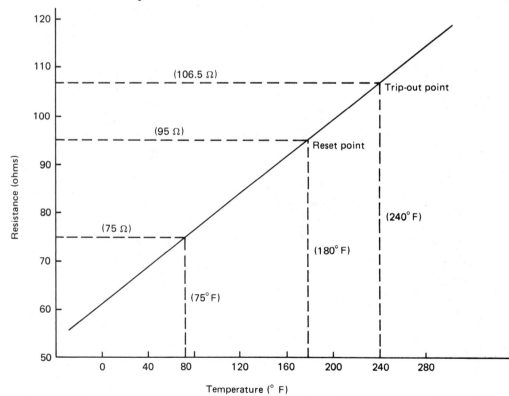

FIGURE 14.19 Graph of the sensor trip-out point and reset point

The module with its wiring connections is shown in Figure 14.20. The module requires a power supply, the sensor connections, the control circuit contacts, and the manual reset connections. Figure 14.21 shows the connections that are required between the module and other components in the system.

FIGURE 14.20 Wiring connections of a module

FIGURE 14.21 Control circuit of a unit with an overload module *(Reproduced by permission of Carrier Corporation, subsidiary of United Technologies)*

Troubleshooting the overload module and sensors is a basic procedure. It requires only a few simple checks to determine the conditions of the module and the sensors. The first and basic check for the entire system is to determine if the control contacts are open or closed, thus eliminating the total overload protection system. If the control contacts are closed, then the problem is in other components. If the control contacts are open, the problem lies in one of three areas: the module, the sensors, or an actual overload. If it is determined that the motor is in an overload condition, then the motor must be checked to determine its condition. A service mechanic should always check the motor and its condition before making a decision about the module and the sensors.

To check each of the three sensors mounted in the motor, you use an ohmmeter. When checking the sensors, follow this procedure:

1. Disconnect the power to the circuit.
2. Check the maximum test voltage of 6 volts.
3. Do not short the sensors.
4. Do not bypass the fuses if they are used.
5. Check the resistance reading of the sensors:
   a. If the resistance is above 95 ohms, wait for a reset.
   b. If the sensor is open, install a 75-ohm, 2-watt resistor.
   c. If the sensor is good, replace the module.

The basic checks after determining the problem in the module and sensor circuit are simple and easy. If all the sensors are good, the module should be replaced. If one sensor is bad, replace it with a 75-ohm, 2-watt resistor.

A wiring harness can be used to check the module in a more detailed procedure. Figure 14.22 shows a wiring diagram for the wiring harness.

Most compressor manufacturers have developed an alternative method of protecting the compressor if the sensors become faulty. For example, Figure 14.23 shows a compressor with a discharge line thermostat used for protection.

Electronic solid-state components can be easily diagnosed for problems. The service mechanic simply checks the signal being sent back to the control module from the thermistors located in the specific control location.

14.3 ADVANCED ELECTRONIC CONTROLS

Electronic control systems are increasingly popular on large heating and air-conditioning systems, commercial refrigeration systems combined for more than one interior case, and sophisticated, small, energy-efficient heat

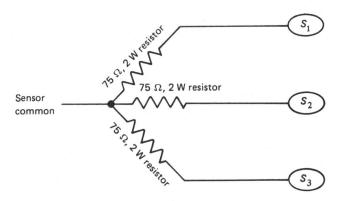

FIGURE 14.22 Wiring diagram for a wiring harness

FIGURE 14.23 Compressor with a discharge line thermostat

pumps. Large heating and air-conditioning systems now use a total solid-state control module for the entire system. Large rooftop units with multizone application especially have gone to the electronic control concept, because electronic controls give higher operating efficiency and better control. Often these systems are incorporated with a variable-air-volume air system, which further complicates control systems because of the varying fan speed. Commercial refrigeration systems that use one condensing unit for several cases, such as those in a large grocery store, are incorporating electronic control for better control of each particular case and for better defrosting of the entire system. Manufacturers of small residential heat pumps now use a total electronic control system to provide better control for the consumer and better operation of the defrost cycle along with higher efficiency.

Many manufacturers of residential heat pumps use a solid-state control system to control the entire operation of the heat pump, including normal operation, operation of supplementary heat, and the defrost cycle. These new heat pumps are supposedly more energy-efficient with better comfort control.

Figure 14.24 shows the control system of a heat pump. The new heat pumps are easy to troubleshoot because most manufacturers have developed a tester to troubleshoot the entire system. Figure 14.25 shows the trouble-shooter used to diagnose problems in one of the modern heat pumps. The tester is accompanied by detailed instructions on its operation.

Large rooftop units are becoming increasingly complicated because of the use of more solid-state control systems. One manufacturer of large roof-top units uses several solid-state control components for the entire system. Large rooftop multizone units usually have many zones controlled from the same basic unit.

Figure 14.26 shows a large multizone rooftop unit installed on a school. This type of unit has a control thermostat for each zone that controls the zone as well as the operation of the entire system. The thermostat sends a signal to a load analyzer shown in Figure 14.27. This device analyzes the signals and determines the operation of the unit. The thermostat, which is

FIGURE 14.24 Control system of a heat pump

FIGURE 14.25 System tester for heat pump

FIGURE 14.26 Large multizone rooftop unit

FIGURE 14.27 Load analyzer

shown in Figure 14.28, also determines the mixture of air that is required by each zone. Figure 14.29 shows the damper motors of a multizone rooftop unit. The damper motor controls the air mixture sent to each zone.

Many rooftop units also utilize a variable-air-volume system that is controlled by a solid-state motor speed-control unit, which is shown in Figure 14.30. Figure 14.31 shows the gas heat operation diagram for this type of unit. A factory service manual is usually required to service this equipment.

FIGURE 14.28 Solid-state thermostat for rooftop unit

FIGURE 14.29 Damper motor setup

FIGURE 14.30 Solid-state motor speed control unit

GAS HEAT OPERATION WITH CONDENSER HEAT ON

GAS HEAT OPERATION WITH CONDENSER HEAT OFF

FIGURE 14.31 Operation diagram of the heat cycle of a rooftop unit *(Redrawn with permission of Lennox Industries, Inc.)*

One of the most popular solid-state control devices is an economizer control package that can be installed as a retrofit on many units that are presently in use in the field. The solid-state electronic economizer receives temperature information from selected sensors, interprets the information, and sends the correct signal to the heating, ventilating, and air-conditioning systems. A labeled flow diagram for the economizer control is shown in Figure 14.32. Figure 14.33 shows the load-sequenced economizer. Figure 14.34 shows several solid-state components used on this type of system. These economizer systems are becoming increasingly popular because of their energy savings.

SUMMARY

The heating, cooling, and refrigeration industry is beginning to use many solid-state components and control systems. Solid-state components and control systems allow better control of temperature, better and faster-acting safety controls, flexibility in speed control of motors, and more efficient defrosting systems for heat pumps and commercial refrigeration equipment. Solid-state controls are used in most phases of the industry, from the large rooftop units to the small split-phase hermetic compressor motors.

Solid-state controls are a vast improvement over the old vacuum tubes. The newer solid-state components are smaller and more durable than the tubes. Thus, manufacturers have been able to incorporate them in the mod-

FIGURE 14.32 Flow diagram for an economizer control unit *(Courtesy of Johnson Controls, Inc.)*

FIGURE 14.33 Load-sequenced economizer

FIGURE 14.34 Solid-state components in the economizer system

ern heating, air-conditioning, and refrigeration control circuits.

The terms "solid state" and "semiconductor" may refer to the same device. Semiconductor devices are very versatile and can provide a wide range of electrical characteristics, making them easy to use in the control systems in the industry. Solid-state devices have one disadvantage: They cannot stand voltage surges. Therefore, sometimes it is necessary to protect solid-state devices from these surges.

The diode and the rectifier are similar solid-state devices. They allow voltage to flow in one direction more easily than in the other direction. The diode is rated at 1 ampere or less, while the rectifier can handle larger loads.

Transistors are used to amplify signals from a thermistor to establish a strong enough signal to cause an action on a relay or contactor. The thermistor is a device that varies its resistance with the temperature.

Many solid-state modules are used in the industry that control one function of the equipment. The simplest one-function control is a solid-state device that acts as a starting relay for small split-phase hermetic compressor motors. A solid-state time-delay device has replaced many time-delay relays. Motor protection has become vastly improved because of the use of solid-state safety systems. Many defrosting systems on heat pumps have also begun to use solid-state devices to achieve better and more efficient defrosting of the outdoor coil in winter operation.

Large heating and air-conditioning systems have also gone to solid-state control systems. Such systems are much more sophisticated than the one-function control modules. Larger systems use solid-state control as a total system control for the entire system.

Troubleshooting solid-state devices is relatively simple. The mechanic must know only what signal must be carried to the module to cause the correct action of the system. The mechanic must also learn to troubleshoot the module and the thermistors that send the signal. Many of the solid-state control systems have accompanying testers with which to check the systems, simplifying the task for the mechanic. The larger systems, however, often require a factory service manual.

QUESTIONS

1. What is the difference between a vacuum tube and a transistor?

2. What is a semiconductor?

3. True or false. The conductivity of materials plays an important part in the action of semiconductors.

4. What is a thermistor?

5. What materials are used to make semiconductors?

6. What is a diode?

7. What is a rectifier?

8. Draw a diagram of a transistor.

9. Briefly explain the rectifier circuit shown in Figure 14.5.

10. What is a transistor?

11. What is the purpose of an anti-short-cycling device?

12. Explain the operation of a solid-state start-assist component used on a PSC motor.

13. Explain the operation of a solid-state starting relay used on a fractional-horsepower split-phase hermetic compressor motor.

14. How have the new solid-state devices affected the control circuits in heat pumps?

15. What is a sensor?

16. Where are sensors located in respect to motor protection?

17. What is the trip-out and reset point in reference to a solid-state motor protector?

18. What is the proper procedure to troubleshoot a solid-state motor protector?

19. How has the emergence of solid-state controls affected large multizone rooftop units?

20. True or false. The repair of solid-state modules will be accomplished by repairing the module.

Lab Section

PREFACE

A refrigeration, heating, and air-conditioning mechanic or technician diagnoses problems in and repairs refrigeration, heating, and air-conditioning systems. Of course, an instructor cannot cover every possible system problem that a technician will face. But if the student has the proper attitude and a broad foundation of knowledge, he or she can be a successful refrigeration, heating, and air-conditioning technician. This lab section attempts to unite the student's knowledge with practical diagnosis and repairs. The basics of system operation and the function of each system component are emphasized.

Reading Electrical Meters

PURPOSE

Because refrigeration, heating, and air-conditioning systems use electrical controls to provide the correct temperature or condition, you must be able to diagnose them. Your primary tool will be the electrical meter, so you must know how to use this test equipment.

REFERENCES

Chapter 3

MATERIALS

1. Analog volt-ohmmeter
2. Digital volt-ohmmeter
3. Clamp-on ammeter

LABORATORY SEQUENCE

I. You should know the type of electrical reading being made and adjust the meter to the correct range and scale.

II. Reading Voltage: Air-conditioning loads and control circuits require a specific voltage for proper operation. Most control systems need 24 volts and the major load, the compressor, needs 115, 208, 230, or 460 volts. In this section you will read control system voltages and line voltages used by important system loads.

III. Using an analog volt-ohmmeter and a digital volt-ohmmeter, record the voltages applied to the following components.

A. With Supervision

    1. Wall receptacle _____ volts

    2. Compressor _____ volts

    3. Evaporator fan motor _____ volts

    4. Condenser fan motor _____ volts

    5. Transformer output _____ volts

B. Without Supervision

    1. Electric heater _____ volts

    2. Wall receptacle _____ volts

    3. Compressor _____ volts

    4. Compressor _____ volts

    5. Evaporator fan motor _____ volts

    6. Electric motor _____ volts

    7. Transformer output _____ volts

    8. Transformer input _____ volts

    9. Light bulb _____ volts

    10. Furnace _____ volts

C. Have your instructor check the voltage readings.

IV. Using the clamp-on ammeter, record the current draw of the following components.

A. With Supervision

    1. Compressor _____ amps

    2. Refrigerator _____ amps

3. Electric heater _____ amps

4. Evaporator fan motor _____ amps

5. Condenser fan motor _____ amps

B. Without Supervision

1. Compressor _____ amps

2. Compressor _____ amps

3. Electric heater _____ amps

4. Evaporator fan motor _____ amps

5. Electric motor _____ amps

6. Condenser fan motor _____ amps

7. Electrical panel _____ amps

8. Transformer output _____ amps

9. Pump motor _____ amps

10. Large compressor _____ amps

C. Have your instructor check the amperage readings.

V. Using an analog volt-ohmmeter and a digital volt-ohmmeter, record the resistance of the following components.

A. With Supervision

1. Heater _____ ohms

2. Electric motor _____ ohms

3. Electric heater _____ ohms

4. Evaporator fan motor _____ ohms

5. Resistor _____ ohms

B. Without Supervision

1. Compressor motor _____ ohms

2. Compressor motor _____ ohms

3. Contactor coil _____ ohms

4. Relay coil _____ ohms

5. Transformer primary _____ ohms

6. Transformer secondary _____ ohms

7. Electric heater _____ ohms

8. Fuse _____ ohms

9. Thermostat _____ ohms

10. Electric motor _____ ohms

C. Have your instructor check the resistance readings.

VI. Record the following circuit characteristics.

1. Compressor _____ volts _____ amps _____ ohms

2. Electric heater _____ volts _____ amps _____ ohms

3. Evaporator fan motor _____ volts _____ amps _____ ohms

4. Condenser fan motor _____ volts _____ amps _____ ohms

5. Electric motor _____ volts _____ amps _____ ohms

A. Have your instructor check the readings.

B. Answer the following questions.

1. If you are checking the voltage available to an air-conditioning unit but have no idea what it could be, what scale should you use?
   a. highest voltage scale
   b. lowest voltage scale

   c. the mid-voltage scale

   d. the voltage scale you think is correct

2. To check a resistance reading of approximately 26,000 ohms, which scale should you use?
   a. r × 1
   b. r × 10
   c. r × 100
   d. r × 1000

3. A mechanic checks the current draw of a small electric motor but cannot accurately determine the low reading. The mechanic decides to loop the conductor supplying the load through the jaws of a clamp-on ammeter four times. The reading is 10 amps. What is the true current draw of the motor?
   a. 1 amp
   b. 2 amps
   c. 2½ amps
   d. 3½ amps

4. At what point on the scale does an electrical meter read most accurately?
   a. lowest point
   b. midpoint
   c. highest point
   d. any point is accurate

5. An advantage of using a digital meter is that _____.
   a. no interpolation of the meter scale is required
   b. no scale setting is required
   c. the meter is foolproof
   d. higher voltages can be read easily

Series and Parallel Circuits

PURPOSE

The major problems encountered when troubleshooting heating, air-conditioning, or refrigeration systems are electrical. Series circuits are generally used as control systems, so if one control opens, the load is removed from the circuit. Parallel circuits are used to supply power to the system loads. Series-parallel circuits, which combine series circuits and parallel circuits, are used in many control systems.

REFERENCES

Chapter 2

MATERIALS

1. Two cleat receptacles (light bulb sockets)
2. Test cord with 110-volt plug
3. Wire
4. Wire terminals
5. Three single-pole switches
6. Two cone heaters (600 watts)
7. One volt-ohmmeter
8. One clamp-on ammeter
9. One 12″ × 12″ piece of plywood
10. Assorted wood screws
11. One terminal board
12. Two 60-watt light bulbs
13. One 5-ampere plug fuse

LABORATORY SEQUENCE

A. Series Circuit Connections and Calculations

1. Side by side in the middle of the plywood, mount two cleat receptacles approximately 6 inches apart. Mount the terminal board 4 inches above the cleat receptacles. The plywood boards should resemble Lab Figure 1. Mark the terminals of the cleat receptacles and terminal board as shown in the figure.

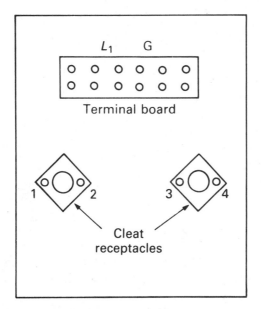

LAB FIGURE 1

2. Wire the board as in Lab Figure 2. The two cleat receptacles form a series circuit, shown schematically in Lab Figure 3.
3. Connect the test cord with the 110-volt plug to terminals L1 and G of the terminal board.
4. Have the instructor check your series circuit.
5. Install the two cone heaters in the cleat receptacles.

LAB FIGURE 2   LAB FIGURE 3

6. Plug the circuit into a 110-volt receptacle. Measure and record in the following table the electrical readings requested.

	Measured	Calculated
a. Voltage at L1 and G		
b. Voltage at 1 and 2		
c. Voltage at 3 and 4		
d. Voltage at L1 and 2		
e. Total resistance		
f. Total amperage		

7. Calculate the requested information and record it in the table.
8. Unplug the circuit from the 110-volt receptacle.
9. Have your instructor check the measured and calculated electrical values.
10. Disconnect the series circuit from the terminal board.
11. Draw a wiring diagram of a series circuit with one switch in series with the cleat receptacles. Use your diagram to wire the circuit and connect it to L1 and G on the terminal board. Remove the cone heaters from the series circuit and install the two light bulbs.
12. Have your instructor check your wiring diagram and wired circuit.
13. Plug the series circuit into a 110-volt receptacle and close the switch. Notice the brightness of the light bulbs and record the voltage applied to each light bulb. Light bulb 1 (terminals 1 and 2): _____ volts; light bulb 2 (terminals 3 and 4): _____ volts.
14. Open the switch and disconnect the circuit from the 110-volt source.
15. Remove the light bulb from receptacle 2 and install the 5-ampere plug fuse.
16. Connect the circuit to the 110-volt source and close the switch. Notice the brightness of the light bulb and record the voltage applied to each receptacle. Light bulb 1 (terminals 1 and 2): _____ volts; cleat receptacle 2 (terminals 3 and 4): _____ volts.
17. Open the switch and remove the circuit from the 110-volt source.
18. Unplug the circuit from the 110-volt receptacle.
19. Remove the wiring from the circuit on the plywood board except for the 110-volt test cord connected to L1 and G on the terminal board.
20. Questions
    a. Why are the light bulbs dimmer in step 12 than the single light bulb in step 15?
    b. Why is the voltage different in step 12 than in step 15?

B. Parallel Circuit Connections and Calculations

1. Wire the board as in Lab Figure 4. The two cleat receptacles are wired in parallel, as shown in Lab Figure 5.
2. Have the instructor check your parallel circuit.
3. Install the two cone heaters in the cleat receptacles.
4. Plug the circuit into a 110-volt receptacle. Measure and record in the following table the electrical readings requested.

		Measured	Calculated
a.	Voltage at L1 and G	_____	_____
b.	Voltage at 1 and 2	_____	_____
c.	Voltage at 3 and 4	_____	_____
d.	Amperage L1 and 1	_____	_____
e.	Amperage 1 and 3	_____	_____
f.	Total amperage	_____	_____
g.	Resistance 1 and 2	_____	_____
h.	Resistance 3 and 4	_____	_____
i.	Total resistance	_____	_____

LAB FIGURE 4

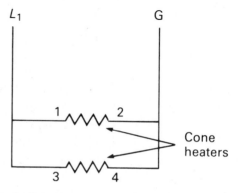

LAB FIGURE 5

5. Unplug the circuit from the 110-volt receptacle.
6. Calculate the requested information and record it in the table.
7. Have the instructor check the measured and calculated values.
8. Disconnect the parallel circuit from the terminal board.

C. Series-Parallel Circuits

1. The series-parallel circuit is a combination of the series and parallel circuits. The electrical circuitry of most air conditioning equipment is composed of series-parallel circuits. An example of a series-parallel circuit is shown in Figure 2.10 of the text.
2. Draw a schematic diagram of a series-parallel circuit to meet each of the following specifications.
   a. *Circuit A:* Control two light bulbs connected in parallel by a switch connected in series with both bulbs.
   b. *Circuit B:* Connect two light bulbs in parallel and control each bulb by a switch in series with each bulb.
   c. *Circuit C:* Connect a switch in series with the parallel combination of both light bulbs, each in series with a switch.
   d. Have your instructor check each schematic.
   e. Properly wire circuits A, B, and C. Have the instructor check each circuit before applying 110 volts and operating the switches.
   f. Remove all wiring and equipment from the plywood and return it to its proper place.

D. Answer the following questions.

1. Two 110-volt light bulbs connected in series with 110 volts would _____.
   a. burn out immediately
   b. burn correctly
   c. burn dimly
   d. do none of the above

2. Two 110-volt light bulbs connected in parallel with 110 volts would _____.
   a. burn out immediately
   b. burn correctly
   c. burn dimly
   d. do none of the above

3. A thermostat, high-pressure switch, and low-pressure switch are connected in series with a small compressor. If the low-pressure switch opened, the compressor would _____.
   a. continue to run
   b. stop immediately
   c. stop if all three safety switches opened
   d. do none of the above

4. Four light bulbs of the same wattage are connected in parallel with 110 volts. The voltage at each bulb is _____.
   a. 110
   b. 55
   c. 27.5
   d. 0

5. Four light bulbs of the same wattage are connected in series with 440 volts. The voltage at each bulb is _____.
   a. 440
   b. 220
   c. 110
   d. 55

Electrical Symbols Used in the HVAC Industry

PURPOSE

You must know how to read schematic diagrams of control systems to troubleshoot electrical systems in HVAC equipment. Therefore, you must be able to recognize the symbols of the electrical components in the control systems. Although most symbols used in the industry are standard, always check the legend of the wiring diagram for the names of the symbols being used; you will find some differences.

REFERENCES

Sections 4.1 through 4.6

MATERIALS

1. Selection of common electrical components
2. Packaged air conditioner
3. Heat pump

LABORATORY SEQUENCE

A. Identify the symbols on the following page.

1.

2.

3.

4.

5.

6.

CH

7.

8.

9.

10.

COMP

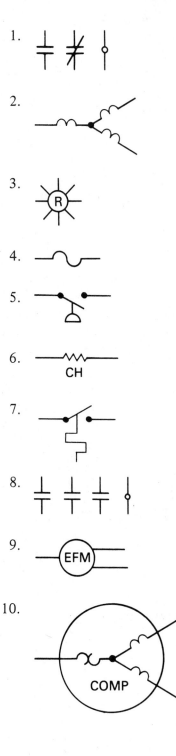

11.

12.

13.

14.

15.

16.

17.

18.

19.

20.

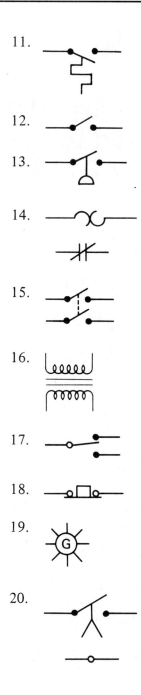

B. Draw the symbols for the following components.

1. Fuse
2. Red signal light
3. Compressor motor
4. Three-speed indoor fan motor
5. Reversing valve solenoid
6. Crankcase heater
7. Heating thermostat
8. Cooling thermostat
9. Three-phase compressor motor
10. Relay with one NO contact and one NC contact
11. Magnetic overload
12. Transformer
13. Capacitor
14. Low-pressure switch (closes on rise in pressure)
15. High-pressure switch (opens on rise in pressure)
16. Thermal line break overload
17. Hot gas solenoid
18. Supplementary heater
19. Three-pole contactor
20. Magnetic starter

C. Draw symbols for electrical components.

1. Obtain 10 different electrical components.
2. Draw the symbols for each component.
3. Have your instructor check the symbols.

D. Locating Components from Symbols

1. Your instructor will assign a unit on which to identify the following components:
    a. Condenser fan motor
    b. Evaporator fan motor
    c. Compressor motor
    d. Transformer
    e. Thermostat
    f. Capacitor

2. Locate the following components on the heat pump assigned by your instructor:

a. Reversing valve solenoid
b. Electrical resistance heater
c. Low-pressure switch
d. Crankcase heater
e. Defrost relay
f. Contactor
g. Thermal overload for electric heaters
h. Thermostat

E. Answer the following questions.

1. How many motors are shown in Figure 4.38?
   a. one
   b. two
   c. three
   d. four

2. Component OFC1 in Figure 13.08 is a _____.
   a. two-pole contactor
   b. three-pole contactor
   c. four-pole contactor
   d. two-pole relay

3. The defrost relay in Figure 13.23 is a _____.
   a. relay with two NO contacts and one NC contact
   b. relay with two NC contacts and one NO contact
   c. relay with three NO contacts
   d. relay with one NO contact

4. How many heating thermostats are shown in Figure 13.23?
   a. one
   b. two
   c. three
   d. four

5. The indoor fan motor in Figure 13.23 is a _____.
   a. one-speed motor
   b. two-speed motor
   c. three-speed motor
   d. four-speed motor

Reading Schematic Diagrams

PURPOSE

A schematic diagram shows a systematic layout of a control system. It can tell you how, when, and why a system works as it does. Reading the diagrams is easy if you take one circuit at a time instead of trying to follow the entire diagram at once. All schematic diagrams are broken down into basic circuits, and each circuit usually contains one load.

REFERENCES

Sections 4.8 through 4.11

LABORATORY SEQUENCE

A. Answer the following questions using Figure 4.38.

    1. The HP and LP switches are connected in _____ with the compressor motor.
       a. series
       b. parallel

    2. The compressor motor is protected by an _____ overload.
       a. internal
       b. external

    3. The condenser fan motor is protected by an _____ overload.
       a. internal
       b. external

4. The contactor is controlled by _____.
   a. the disconnect switch
   b. the high-pressure switch
   c. the cooling thermostat
   d. none of the above

5. With the fan switch in the "on" position, the indoor fan will _____.
   a. run when the cooling thermostat is closed only
   b. run when the temperature exceeds 95° F
   c. run continuously
   d. do none of the above

6. The primary voltage of the transformer is _____.
   a. 230 volt
   b. 208 volt
   c. 115 volt
   d. 208/230 volt

7. The compressor and condenser fan motors start _____.
   a. at separate times
   b. at the same time

8. The compressor and condenser fan motors are connected in _____.
   a. series
   b. parallel

9. Write an operational sequence for the condensing unit.

10. Have your instructor check your answers to questions 1–9.

B. Answer the following questions using Figure 12.11.

1. The control circuit of the unit is a _____.
   a. line voltage circuit
   b. low-voltage circuit

2. The thermostat cycles _____.
   a. the compressor
   b. the indoor fan motor
   c. the compressor and indoor fan motor
   d. none of the above

3. The holding relay and time assembly work together as _____.

a. an operating control
b. an anti-short-cycling function
c. an indoor fan delay function
d. none of the above

4. The outdoor fan relay is energized by the thermostat starting the
_____.
a. compressor
b. indoor fan motor
c. outdoor fan motor
d. crankcase heater

5. The contactor coil is energized by the closing of _____.
a. the thermostat
b. the high-pressure switch
c. the holding relay contacts (3 and 1)
d. the timer contacts (A and A1)
e. all of the above

6. If the unit is calling for cooling, the holding relay coil is energized
by the _____.
a. HR contacts (3 and 1)
b. HR contacts (5 and 4)
c. timer contacts (A and A1)
d. OFR contacts

7. Write an operational sequence for the air-cooled packaged unit.

8. Have your instructor check your answers to questions 1–7.

C. Answer the following questions using Figure 13.12.

1. The control relay is energized by _____.
a. opening the HPS
b. closing IT
c. closing the indoor thermostat
d. the transformer

2. The crankcase heater is energized on the _____.
a. "on" cycle
b. "off" cycle

3. The control relay has _____.

a. two NO contacts
b. two NC contacts
c. one NO contact and one NC contact
d. two NO contacts and one NC contact

4. How many safety devices are located in the contactor coil circuit?
a. 2
b. 3
c. 4
d. 5

5. Which of the following controls will prevent the CR from energizing?
a. HP
b. LP
c. IT
d. OL1

6. When the CR coil is energized after an "off" cycle of at least 5 minutes, what major components of the system will be energized?
a. compressor
b. condenser fan motor
c. crankcase heater
d. contactor and condenser fan motor

7. Explain the operation of the TM, TM switches (A, A1, and A2; B, B1, and B2), and the HR.

8. Write an operational sequence for the condensing unit.

9. Have your instructor check your answers to questions 1–8.

D. Answer the following questions using Figure 13.23.

1. This control system is for a _____.
a. gas furnace
b. packaged air conditioner
c. water chiller
d. heat pump

2. The thermostat for this unit is _____.
a. one-stage cooling, one-stage heating
b. two-stage cooling, one-stage heating
c. two-stage cooling, two-stage heating
d. one-stage cooling, two-stage heating

3. The OFM is de-energized when _____.
   a. the DFR is de-energized
   b. the CR is de-energized
   c. the IFR is energized
   d. none of the above

4. When the DFR coil is energized in the normal heating cycle operation, which controls will be de-energized?
   a. RVS
   b. OFM
   c. IFM
   d. a and b

5. The DFT on closing will _____ a defrost cycle.
   a. initiate
   b. terminate

6. Write an operational sequence for the heat pump.

7. Have your instructor check your answers to questions 1–6.

E. Answer the following questions using Figure 4.47.

1. The transformer is located in _____.
   a. the outdoor unit
   b. the indoor unit
   c. the disconnect 1
   d. none of the above

2. What does the first-stage cooling thermostat energize?
   a. MS
   b. F
   c. SC
   d. DFT

3. What does the second-stage cooling thermostat energize?
   a. MS
   b. F
   c. SC
   d. DFT
   e. a and b

4. What does the first-stage heating thermostat energize?

   a. SC
   b. MS
   c. F
   d. AH
   e. b and c

5. What does the second-stage heating thermostat energize?
   a. SC
   b. MS
   c. F
   d. AH
   e. c and d

6. When the RHS-1 and RHS-2 switches are moved to the emergency heat position, which components are energized?
   a. AH
   b. MS
   c. SC
   d. compressor

7. CR-A is energized by the _____.
   a. MS
   b. SC
   c. F
   d. closing of disconnect 1

8. On the defrost cycle the outdoor fan motor is _____.
   a. energized
   b. de-energized

9. The defrost cycle is terminated by _____.
   a. DFT contacts (3 and 5)
   b. DT
   c. DFT contacts (3 and 5) or DT
   d. ODS

10. Write an operational sequence for the heat pump.

11. Have your instructor check your answers to questions 1–10.

F. Draw a schematic wiring diagram to meet the following specifications.

   1. Unit Type: 3-ton packaged air conditioner

2. Compressor Type: PSC motor with internal overload

3. Evaporator Fan Motor: 2-speed motor

4. Condenser Fan Motor: 1-speed motor

5. Control System: 24 volt

6. Safety Controls: Low- and high-pressure switches

G. Answer the following questions.

1. A schematic diagram shows the components in _____.
   a. an energized position
   b. a de-energized position

2. A schematic diagram is _____.
   a. an exact picture of a control panel with connecting wiring
   b. an installation diagram
   c. a circuit-by-circuit layout of the control system
   d. none of the above

3. A schematic diagram is most important for _____.
   a. servicing air-conditioning control systems
   b. locating components in the control panel
   c. installing air-conditioning equipment
   d. homeowner

4. Most loads shown in schematic diagrams are connected in _____.
   a. series
   b. parallel

5. Most switches shown in schematic diagrams are connected in _____.
   a. series
   b. parallel

Sizing Wire for an Air-Conditioning Condensing Unit

PURPOSE

It is most important to size correctly the power wire installed on the unit. An inadequate power supply can damage the equipment. Use the equipment installation instructions or the National Electric Code to size or check the conductors supplying power to the unit. You must also follow local codes and ordinances and the National Electrical Code when making the connections at the electrical panel and on the equipment.

REFERENCE

Chapter 6

MATERIALS

1. National Electrical Code
2. Installation instructions of unit assigned by instructor
3. Air-conditioning condensing unit
4. Wire and supplies to install power wiring to unit

LABORATORY SEQUENCE

A. Sizing Wire for the Condensing Unit Using the Installation Instructions

1. See instructor for assignment of air-conditioning condensing unit.

2. Read the section in the installation instructions on wiring the unit.

3. Follow the procedure described in the installation instructions to determine the correct wire size for the assigned unit. Take note of the

distance between the condensing unit and the supply electrical panel. (**Note:** The type of insulation on the conductor plays an important part in its current-carrying ability.)

4. Complete data sheet A.

Data Sheet A

Condensing Unit Model Number: _____

Condensing Unit Serial Number: _____

Distance from Power Supply to Unit: _____

Size of Wire to be Installed: _____

List Material Required to Complete Installation:

_____

_____

_____

_____

5. Have instructor check wire sizing and material list for assigned condensing unit.

B. Sizing Wire for the Condensing Unit Using the National Electrical Code

1. See instructor for assignment of air-conditioning condensing unit.

2. Using the National Electrical Code, size the power wiring connecting the assigned condensing unit to the electrical power panel. (**Note:** Refer to paragraph 6.1 if you need help.)

3. Calculate the voltage drop in the conductor from the unit to the panel.

4. Complete data sheet B.

Data Sheet B

Condensing Unit Model Number: _____

Condensing Unit Serial Number: _____

Distance from Power Supply to Unit: _____

Size of Wire to be Installed: _____

List Material Required to Complete Installation:

_____

_____

_____

_____

5. Have instructor check wire sizing and material list for assigned unit.

C. Answer the following questions.

1. According to the National Electrical Code, what size wire has an ampacity of 65 amps with THWN insulation?
   a. No. 10
   b. No. 8
   c. No. 6
   d. No. 4

2. What is the size and insulation type of the conductor to supply a condensing unit whose compressor draws 36 amps and whose condenser fan motor draws 5 amps?
   a. No. 6 TW
   b. No. 8 TW
   c. No. 10 THWN
   d. No. 12 THHN

3. The installation mechanic should always check the voltage drop of _____.
   a. short power circuits (under 50 feet)
   b. long power circuits (over 200 feet)
   c. each circuit
   d. none of the above

4. When in doubt about wiring procedures, and no local codes are in force, always _____.
   a. follow the National Electrical Code
   b. supply No. 8 TW to condensing units over 5 hp
   c. use Ohm's law
   d. none of the above

5. The best overall conductor for power wiring circuits is _____.
   a. aluminum
   b. copper
   c. nickel-coated copper
   d. silver

Capacitors

PURPOSE

Most single-phase equipment has at least one and possibly several capacitors to help start electric motors. In some motors, the capacitor (run capacitor) remains in the starting circuit for good running efficiency; in others, the capacitor (start capacitor) is removed from the circuit after starting the motor. Some type of relay or centrifugal switch is generally used to remove the capacitor from the circuit. The run capacitor is oil filled to help dissipate heat, but the start capacitor has no way to dissipate heat. You must be able to determine the condition of capacitors and understand the rules that apply when replacing a capacitor with one that is not an exact duplicate. In this lab exercise you will check capacitors in motor circuits and determine correct replacements.

REFERENCES

Section 7.5

MATERIALS

1. 20 numbered capacitors
2. Volt-ohmmeter
3. Capacitor testers
4. 2-watt 15,000-ohm resistor

LABORATORY SEQUENCE

A. Determining the Size and Type of Capacitor

1. Obtain 20 capacitors from your instructor.

2. Using a 2-watt, 15,000-ohm resistor, short between the terminals of the capacitors to discharge any capacitor that may be charged.

(**Note:** Capacitors often remain charged even while sitting on the shelf. Be safe.)

3. Record the types, voltages, and microfarad ratings of the capacitors numbered from 1 to 10.

4. Have instructor check your work sheet.

B. Checking the Condition of Capacitors

1. Check the 20 capacitors with a capacitor tester and a volt-ohmmeter.

2. Record each capacitor's condition as "good," "shorted," or "open."

3. Have instructor check your work sheet.

C. Answer the following questions, using the capacitor rules in paragraph 7.5 of the text.

1. The total capacitance of two 150-μF capacitors connected in series is _____.
   a. 25 μF
   b. 50 μF
   c. 75 μF
   d. 100 μF

2. The total capacitance of a 15-μF capacitor and a 45-μF capacitor connected in parallel is _____.
   a. 11 μF
   b. 30 μF
   c. 60 μF
   d. 675 μF

3. A 210-μF, 250-volt start capacitor can be replaced by which start capacitor?
   a. 150-μF, 250-volt
   b. 210-μF, 330-volt
   c. 285-μF, 250-volt
   d. 210-μF, 150-volt

4. A 150-μF, 330-volt capacitor can be replaced by which start capacitor?
   a. 150-μF, 660-volt
   b. 190-μF, 330-volt

   c. 190-μF, 150-volt
   d. 150-μF, 250-volt

5. A 30-μF, 330-volt run capacitor can be replaced by which run capacitor?
   a. two 15-μF, 330-volt run capacitors connected in parallel
   b. 20-μF, 330-volt
   c. 25-μF, 330-volt
   d. 40-μF, 330-volt

6. A 5-μF, 440-volt capacitor can be replaced by which run capacitor?
   a. 4-μF, 440-volt
   b. 5-μF, 250-volt
   c. 5-μF, 370-volt
   d. 5-μF, 550-volt

7. Have your instructor check your answers to questions 1–6.

D. Answer the following questions.

1. In an electric motor a start capacitor increases _____.
   a. the starting torque
   b. the running efficiency
   c. the horsepower
   d. none of the above

2. The run capacitor is designed to _____.
   a. remain in the electrical circuit
   b. be removed from the electrical circuit
   c. be removed from the circuit only when the motor is running at slow speeds
   d. do none of the above

3. A run capacitor is physically larger than a start capacitor because
   _____.
   a. it has a greater microfarad rating
   b. it has a greater voltage
   c. it dissipates more heat
   d. none of the above

4. A volt-ohmmeter indicates a good capacitor when the needle moves
   to _____.
   a. 0 and remains
   b. infinite reading and remains

c. same measurable resistance and remains
d. 0 and falls back toward the infinite reading

5. Which of the following conditions indicates a faulty capacitor?
   a. open
   b. shorted
   c. a and b
   d. none of the above

Electric Motors

PURPOSE

Electric motors are the most important loads in heating, air-conditioning, or refrigeration systems. You must be able to recognize different types of motors in order to troubleshoot and repair them. The major types of motors used in the industry are split-phase, shaded-pole, permanent split-capacitor, capacitor-start capacitor-run, and three-phase. This lab will help you recognize, troubleshoot, and replace the various types of electric motors. Starting components, such as start relays and switches, are also briefly covered.

REFERENCES

Chapters 7 and 8

MATERIALS

1. Shaded-pole motor
2. Permanent split-capacitor motor
3. Split-phase motor
4. Capacitor-start capacitor-run motor
5. Three-phase motor
6. 10 motors to identify and troubleshoot
7. Volt-ohmmeter
8. Clamp-on ammeter
9. Miscellaneous hand tools

LABORATORY SEQUENCE

A. Identifying the Types of Electric Motors

1. Obtain 10 electric motors from your instructor.

2. Identify the 10 motors and record each type in the following chart. (**Note:** Several things can help you identify the motors: (1) Examine the motor and nameplate. (2) Shaded-pole motors have a shading coil that can easily be identified. (3) Permanent split-capacitor motors can be identified because their windings look like a maze of wires. (4) The split-phase and capacitor-start capacitor-run motors can have an open design and be mounted on a frame, or they can be sealed hermetic compressor motors. The open motor can usually be identified by the capacitor on the top of the motor. The hermetic compressor motor is almost impossible to identify without the manufacturer's literature. Capacitor-start capacitor-run motors are used almost exclusively in hermetic compressors. (5) Three-phase motors are identified by their nameplates. If you have trouble identifying the motors, refer to the references.)

1. _____

2. _____

3. _____

4. _____

5. _____

6. _____

7. _____

8. _____

9. _____

10. _____

3. Have your instructor verify your identification of these motors.

B. Shaded-Pole Electric Motors

1. Obtain a shaded-pole motor from your instructor.

2. Check the windings of the motor with a volt-ohmmeter, record the resistance readings, and draw a schematic diagram.

3. Wire the shaded-pole motor, making certain it is wired to the correct power source.

4. Have the instructor check your wiring and your diagram.

5. Operate the motor and record its current draw.

6. Locate a shaded-pole motor on a piece of equipment in the lab.

C. Permanent Split-Capacitor Motor

1. Obtain a permanent split-capacitor motor from your instructor.

2. Check the windings of the motor with a volt-ohmmeter, record the resistance readings, and draw a schematic diagram.

3. Have instructor check your wiring diagram.

4. Wire the permanent split-capacitor motor with the correct capacitor, making certain the motor is wired to the correct power source.

5. Have instructor check your wiring.

6. Operate the motor and record its current draw.

7. Locate a permanent split-capacitor motor on a piece of equipment in the lab. Examine a permanent split-capacitor hermetic compressor motor on an air-conditioning unit in the lab.

D. Split-Phase or Capacitor-Start Open-Type Motor

1. Obtain a split-phase or capacitor-start open-type motor from your instructor.

2. Check the windings of the motor with a volt-ohmmeter, record the resistance readings, and draw a schematic diagram.

3. Wire the motor, making certain the motor is wired to the correct power source.

4. Have instructor check your wiring.

5. Operate the motor and record its current draw.

6. Locate a split-phase motor on a piece of equipment in the lab.

E. Capacitor-Start Capacitor-Run Hermetic Compressor Motor

1. Obtain a capacitor-start capacitor-run hermetic compressor motor from your instructor.

2. Check the windings of the motor with a volt-ohmmeter, record the resistance readings, and draw a schematic diagram (include the potential relay).

3. Have instructor check your wiring diagram.

4. Wire the motor and potential relay with the necessary capacitors, making certain the motor is wired to the correct power source.

5. Have instructor check your wiring.

6. Operate the motor and record its current draw.

7. Locate a capacitor-start–capacitor-run hermetic compressor motor on an air-conditioning unit in the lab.

F. Three-Phase Motor

1. Obtain a three-phase motor from your instructor.

2. Check the windings of the motor with a volt-ohmmeter, record the resistance readings, and draw a schematic diagram.

3. Have instructor check your wiring diagram.

4. Wire the motor, making certain it is wired to the correct power source.

5. Have instructor check your wiring.

6. Operate the motor and record its current draw.

7. Reverse the direction of the motor by switching any two wires.

8. Locate a three-phase motor on a commercial air conditioner in the shop.

G. Testing Electric Motors

1. Obtain 10 motors to troubleshoot from your instructor.

2. Troubleshoot the motors and record the condition and type of each motor in the following chart.

1. _____

2. _____

3. _____

4. _____

5. _____

6. _____

7. _____

8. _____

9. _____

10. _____

3. Have your instructor check your findings.

H. Answer the following questions.

1. Which motor has the highest starting torque?
   a. shaded-pole
   b. permanent split-capacitor
   c. split-phase
   d. three-phase

2. Bad motor windings could be labeled _____.
   a. shorted
   b. grounded
   c. open
   d. all of the above

3. Starting relays are used in hermetic compressor motors to _____.
   a. drop the starting winding out of the circuit
   b. drop the starting winding and components out of the circuit
   c. protect the starting winding
   d. do all of the above

4. Shaded-pole motors are used on _____.
   a. small fans
   b. large centrifugal fans
   c. compressor motors
   d. all of the above

5. A three-phase motor is normally reversed by _____.
   a. reversing any two power leads
   b. rewinding the motor
   c. changing from a star to a delta winding
   d. adding three capacitors

Transformers

PURPOSE

Transformers are used primarily in the industry to supply a specific control voltage: 24 volts in residential equipment and 110 volts in commercial and industrial equipment. Transformers are also used for many special applications, such as buck and boost, oil burner ignition, and electronic air filter power supplies. This lab covers the control transformer.

REFERENCES

Sections 10.1 and 11.6

MATERIALS

1. 24-volt control transformers
2. 110-volt control transformers
3. Volt-ohmmeter
4. Terminal boards
5. Basic hand tools

LABORATORY SEQUENCE

A. Troubleshooting Transformers

1. Obtain a selection of 24-volt and 110-volt control transformers from your instructor.

2. Using a volt-ohmmeter, take resistance readings of the primaries and secondaries of six transformers and record the readings. (**Note:** Do not touch the transformer leads when taking the resistance readings. Electrical shock can occur as the test leads are disconnected.)

3. Connect the primary of each transformer to the correct voltage, and

the secondary of each transformer to a 24-volt or a 110-volt load, such as a contactor coil, as shown in Lab Figure 6. Make these connections on a terminal board.

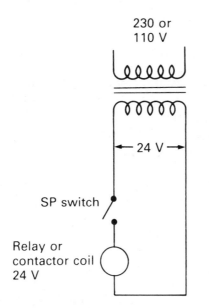

LAB FIGURE 6

4. Using a volt-ohmmeter, read the voltages of the primary and secondary of each transformer with the load connected. Record these readings in the chart. (**Note:** Be careful when taking readings of live circuits.)

5. Determine the condition of each transformer by the readings in your chart.

6. Have instructor check your diagnoses.

B. Installing Transformers

1. Your instructor will assign a unit to replace the transformer.

2. Install the transformer on the assigned unit and record the primary and secondary voltages.

3. Have your instructor check your work.

C. Answer the following questions.

1. In a control circuit a transformer _____.
   a. supplies line voltage
   b. supplies control voltage
   c. boosts control voltage
   d. does none of the above

2. The VA rating of a replacement transformer can be determined by _____.
   a. the load of the control system
   b. the name tag on the faulty compressor
   c. the manufacturer's recommendation
   d. all of the above

3. The resistance reading of a good transformer is _____.
   a. none
   b. infinite
   c. measurable
   d. none of the above

4. What would happen if a 40-VA transformer were installed on a 20-VA load?
   a. The control system will work.
   b. The transformer will burn out.
   c. The control system will work for a short time.
   d. All of the above will occur.

5. Control systems on residential air-conditioning systems are usually _____.
   a. 24 volts
   b. 110 volts
   c. 230 volts
   d. all of the above

Contactors and Relays

PURPOSE

Contactors and relays are used in the industry to stop and start loads. The major difference between a contactor and a relay is the ampacity of the contacts. Common faults of contactors and relays are faulty contacts, defective coils, which are primarily electrical problems, and linkage problems, which are primarily mechanical. In this lab you will troubleshoot and install contactors or relays in circuits that you have designed to certain specifications.

REFERENCES

Sections 9.1, 9.2, and 11.2

MATERIALS

1. Relays and contactors with 24-, 110-, and 230-volt coils
2. 24-volt control transformers
3. Volt-ohmmeter
4. Terminal boards
5. Switches
6. Basic hand tools
7. Cleat receptacles

LABORATORY SEQUENCE

I. Contactor and Relay Operation

    A. 110-Volt Relay Connections

1. Complete Lab Figure 7 so that the single-pole switch controls the relay that turns on the 110-volt light bulb. After you have drawn the pictorial circuits, draw the schematic.

LAB FIGURE 7

2. Have your instructor check both diagrams.

3. Using your diagram, wire the circuit on terminal board.

4. Connect the circuit to a 110-volt power source.

5. Have your instructor check your circuits.

6. Operate the circuit by closing the switch.

7. Record what happens when the switch is closed.

8. Disconnect the power from the circuit and remove the components from the board.

B. 230-Volt Contactor Connections

1. Complete Lab Figure 8 so that the single-pole switch controls the contactor that turns on the 110-volt light bulb. After you have drawn the pictorial circuit, draw the schematic.

2. Have your instructor check both diagrams.

L_1 L_2 G

← 230 V →←—— 110 V ——→

SPST switch   Relay with   110 V
                   230 V coil   light
                                     bulb

LAB FIGURE 8

3. Using your diagram, wire the circuit on the board used in step A3.

4. Connect the circuit to a 230-volt power source. (**Note:** The relay coil requires 230 volts, but the light bulb requires only 110 volts. The voltage from one leg of a 230-volt circuit to ground is 110 volts.) Make your connections on terminal boards.

5. Have your instructor check your circuits.

6. Operate the circuit by closing the switch.

7. Record what happens when the switch is closed.

8. Disconnect the power from the circuit and remove the components from the board.

C. 24-Volt Relay Connections

1. Complete Lab Figure 9 so that the single-pole switch controls the relay that turns on the 110-volt light bulb. After you have drawn the pictorial circuit, draw the schematic.

2. Have your instructor check both diagrams.

3. Using your diagram, wire the circuit on the board used in step A3.

SPST
switch

Relay with
24 V coil

Light
bulb

LAB FIGURE 9

4. Connect the circuit to a 110-volt power source. (**Note:** The relay coil requires 24 volts, which must be provided by the 24-volt transformer, but the light bulb requires only 110 volts. The transformer must also be supplied with the correct power. Check the voltage of the transformer.) Make your connections on terminal boards.

5. Have your instructor check your circuits.

6. Operate the circuit by closing the switch.

7. Record what happens when the switch is closed.

8. Disconnect the power from the circuit and remove the components from the board.

D. 24-Volt Relay, Including Control Circuitry

1. Complete Lab Figure 10 so that the single-pole switch controls

LAB FIGURE 10

the relay that energizes a 110-volt light bulb when the switch is open or closed. After you have drawn the pictorial circuit, draw the schematic.

2. Have your instructor check both diagrams.

3. Using your diagram, wire the circuit on the board used in step A3.

4. Connect the circuit to a 110-volt power source. (**Note:** The circuitry is much the same as in step C4 with the exception of the NC contacts.) Make your connections on terminal boards.

5. Have your instructor check your circuits.

6. Operate the circuit by closing the switch.

7. Record what happens when the switch is closed.

8. Disconnect the power from the circuit and remove the components from the board.

E. Troubleshooting Relays and Contactors

1. Obtain from your instructor a selection of relays and contactors that are not installed on equipment.

2. Troubleshoot each individual relay or contactor and record its condition. The condition of the coil *and* the contacts must be recorded. (**Note:** Relays with an NC set of contacts must be energized to determine the condition of the contacts.)

3. Have your instructor check your list of relay faults.

F. Answer the following questions.

1. The difference between a contactor and a relay is _____.
   a. the coil voltage
   b. the contact arrangement
   c. the ampacity of the contacts
   d. none of the above

2. What does a reading of 50 volts across L1 and T1 on a contactor indicate?
   a. Contacts are open.
   b. Contacts are closed.
   c. Voltage is being lost across contacts.
   d. Resistance of the coil is too high.

3. Common faults in relays and contactors occur in the _____.
   a. coil, contacts, and mechanical linkage
   b. coil, contacts, and wiring
   c. coil, mechanical linkage, and position
   d. coil, shading coil, and contacts

4. If the voltage to a relay coil is correct but the relay remains de-energized, what is the most likely fault?
   a. bad coil
   b. bad contacts
   c. bad mechanical linkage
   d. all of the above

5. If the coil of a contactor shorts and energizes the control system, what would happen?
   a. The transformer would burn out.
   b. The contactor contacts would close.
   c. The contacts would be badly burned.
   d. The load controlled by the contactor would be damaged.

Thermostats

PURPOSE

The basic purpose of a heating, air-conditioning, or refrigeration system is to maintain a certain temperature in a given space. The thermostat is designed to either open or close a set or sets of contacts when the temperature of the space reaches a certain value. The function of the thermostat is basically determined by whether the contacts open or close on an increase or decrease in temperature. In this lab you will install and troubleshoot all types of thermostats, including the subbases.

REFERENCES

Sections 10.2, 10.3, and 11.4

MATERIALS

1. Line voltage thermostats
2. Heating and cooling thermostats
3. Two-stage heating, two-stage cooling thermostats
4. Other components as determined by diagrams
5. Volt-ohmmeter
6. Miscellaneous wiring supplies

LABORATORY SEQUENCE

I. Line Voltage Thermostats

    A. Line Voltage Thermostat Circuitry and Installation

        1. Complete Lab Figure 11 so that the line voltage thermostat energizes the small 110-volt fan motor when the temperature exceeds 85° F. After you have drawn the pictorial circuit, draw the schematic.

LAB FIGURE 11

2. Have your instructor check both diagrams.

3. Using your diagram, wire the circuit on an electrical practice board.

4. Using a terminal board, connect the circuit to a 110-volt power source.

5. Have your instructor check your circuits.

6. Operate the circuit by closing the thermostat.

7. Record what happens when the switch is closed.

8. Disconnect the power from the circuit and remove the components from the board.

B. Troubleshooting Line Voltage Thermostats

1. Obtain three line voltage thermostats from your instructor.

2. Troubleshoot the line voltage thermostats and record the condition of each one. (**Note:** Make certain you know the function of the thermostat.)

3. Have your instructor check your work.

II. Low-Voltage Thermostats

   A. Heating-Cooling, Low-Voltage Thermostat Circuitry and Installation

     1. Complete Lab Figure 12 to form a system to control heating, cooling, and the fan. Light bulbs represent system loads, such as compressors, fan motors, and heating sources. After you have drawn the pictorial circuit, draw a schematic of the control system circuitry.

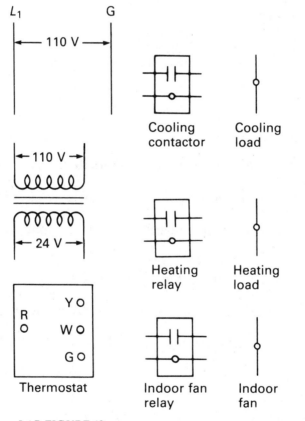

LAB FIGURE 12

     2. Have your instructor check both diagrams.

     3. Using your diagram, wire the control system on an electrical practice board. Read the installation instructions provided with the thermostats.

4. Connect your circuit to the correct power source, using terminal boards.

5. Have your instructor check your circuits.

6. Operate the control system by setting the thermostat for the desired function; check cooling, heating, and fan operation.

7. Record what happens when the thermostat is set for each function.

8. Disconnect the power from the circuit and remove the components from the board.

B. Two-Stage Heating, Two-Stage Cooling, Low-Voltage Thermostat Circuitry and Installation

1. Complete Lab Figure 13 to form a system to control two-stage heating and two-stage cooling with fan. The two-stage cooling two-stage heating means that two parts of heating and cooling are used. If you have any question on staging thermostats, refer to paragraph 10.2 in the text. Light bulbs represent heating and cooling loads. After you have drawn the pictorial circuit, draw a schematic of the control circuitry.

2. Have your instructor check both diagrams.

3. Using your diagram, wire the control system on an electrical practice board. Read the installation instructions furnished with each thermostat.

4. Connect your circuit to the correct power source, using terminal boards.

5. Have your instructor check your circuits.

6. Operate the control system by setting the thermostat for the desired function; check cooling, heating, and fan operation.

7. Record what happens when the thermostat is set for each function.

8. Disconnect the power from the circuit and remove the components from the board.

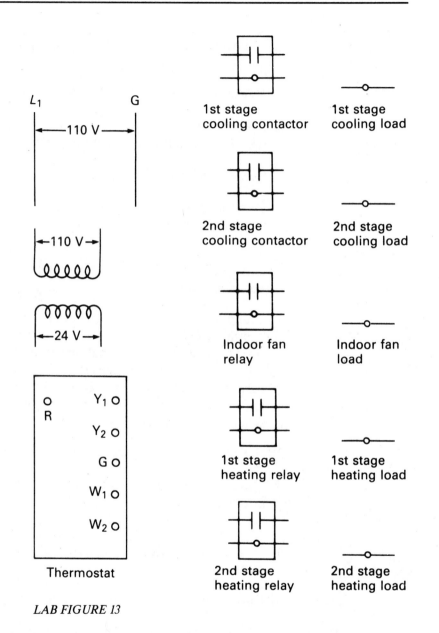

LAB FIGURE 13

C. Troubleshooting Low-Voltage Thermostats

1. Your instructor will assign you to troubleshoot the thermostat on two air-conditioning and heating systems. (**Note:** Troubleshooting a low-voltage thermostat that is not connected to a system is extremely hard because of the subbase and internal control cir-

cuitry. Your results when troubleshooting low-voltage thermostats if you concentrate on the function of the thermostat will reduce the time of troubleshooting. Make certain that the thermostat is correctly set for the function you are checking. You must check all functions to ensure that the thermostat is operating properly.)

2. Troubleshoot the two thermostats and record their faults.

D. Answer the following questions.

1. A heating thermostat _____.
   a. opens on rise in temperature
   b. closes on rise in temperature

2. To effectively troubleshoot thermostats, you must know the thermostat's _____.
   a. type
   b. range
   c. voltage
   d. control system function

3. The mechanic jumpers a line voltage thermostat, but the system does not operate. This indicates _____.
   a. a thermostat fault
   b. a system fault
   c. a subbase fault
   d. none of the above

4. A low-voltage thermostat with mercury bulbs must be mounted _____.
   a. level
   b. with vibration eliminators
   c. exactly 5 feet from floor
   d. none of the above

5. A low-voltage thermostat is usually _____.
   a. a single-function device
   b. a multifunction device
   c. a temperature-sensing device only
   d. none of the above

Pressure Switches

PURPOSE

Pressure switches are used in control circuits as safety switches and operating controls. When a pressure switch is used as a safety switch, it de-energizes a control circuit, usually to the compressor, when an unsafe condition exists. An operating control starts and stops a load for a particular reason, such as head pressure control or temperature control. You can effectively troubleshoot a pressure switch only if you know its function in the system and its pressure setting. There are many varieties of pressure switches; some preset, others adjustable. In this lab exercise you will examine, install, and troubleshoot pressure switches.

REFERENCES

Sections 10.5 and 11.6

MATERIALS

1. Eight pressure switches
2. High-pressure switch (opens on rise)
3. Low-pressure switch (closes on rise)
4. Dual-pressure switch
5. Volt-ohmmeter
6. Miscellaneous wiring supplies

LABORATORY SEQUENCE

A. Identifying Pressure Switch

   1. Obtain eight pressure switches from your instructor.

2. Record the type, function, and setting of each switch.

3. Find and list equipment location of each of the following pressure switches:
   a. Adjustable high-pressure switch (safety)
   b. Fixed high-pressure switch (safety)
   c. Adjustable high-pressure switch (operating control)
   d. Adjustable low-pressure switch (safety)
   e. Fixed low-pressure switch (safety)
   f. Adjustable low-pressure switch (operating control)

4. Have your instructor check your work sheet.

B. Low-Pressure Switch Used As an Operating Control

1. Your instructor will assign a commercial refrigeration system to you to set a low-pressure switch to obtain a particular temperature. He or she will also give you the temperature.

2. Set the pressure switch to obtain the assigned temperature.

3. Operate the unit and record the temperature and off-cycle time.

4. Adjust the pressure switch until the assigned temperature is obtained.

5. Have instructor check your project.

C. High-Pressure Switch Used As an Operating Control

1. From your instructor obtain a high-pressure switch that closes when temperature rises.

2. Your instructor will assign a small air-cooled condensing unit so that you can install the high-pressure switch to maintain the assigned head pressure.

3. Install the high-pressure switch and make all necessary connections (pressure and electrical).

4. Set the pressure switch for the assigned pressure.

5. Have instructor check your pressure connections, wiring connections, and pressure switch setting.

6. Operate the system, observe the actions of the pressure switch, and record the pressures when the condenser fan cuts off and on.

D. Dual-Pressure Switch Used As a Safety Control

1. From your instructor obtain a dual-pressure switch that operates as a safety switch.

2. Your instructor will assign a small air-cooled condensing unit so that you can install the dual-pressure switch as a safety control.

3. Install the dual-pressure switch and make all necessary connections (pressure and electrical). (Use valves to make pressure connections.)

4. Set the pressure switch for the proper safety settings.

5. Have instructor check your pressure connections, wiring connections, and setting of the pressure control.

6. Operate the system and observe the actions of the pressure switch.

7. Block the condenser with a piece of cardboard and operate the system. Record the actions of the system.

8. Close the compressor suction service valve and operate the system. Record the actions of the system.

9. Open the suction service valve and return the system to normal operation.

E. Troubleshooting Pressure Switches

1. Your instructor will assign you two operating refrigeration switches.

2. Troubleshoot the pressure switches on the refrigeration systems. (**Note:** You must understand the function of the pressure switch and then determine the fault.) Record the fault and the pressure settings of the pressure switches as well as the actual cut-in and cut-out of the compressor.

3. Have your instructor check your findings.

F. Answer the following questions.

1. A high-pressure switch used as an operating control to stop and start a condenser fan motor for head pressure control would _____.
   a. open on a rise in pressure
   b. close on a rise in pressure

2. A low-pressure switch used as an operating control could _____.
   a. stop and start a compressor
   b. stop a condenser fan motor
   c. start a crankcase heater
   d. do none of the above

3. When troubleshooting a pressure switch, you must know _____.
   a. refrigeration system pressures
   b. pressure switch function
   c. pressure switch setting
   d. all of the above

4. The position of the contacts in a pressure switch is determined by a _____.
   a. pressure test
   b. visual observation
   c. voltage test
   d. contactor test

5. A dual-pressure switch could be used for _____.
   a. safety control
   b. high-pressure safety control; low-pressure operating control
   c. low-pressure safety control; high-pressure operating control
   d. none of the above

Overloads

PURPOSE

Motors are usually protected by an overload. Overloads come in many types, from in-line overloads for fractional horsepower hermetics to electronic overloads used on large compressors. You must be able to troubleshoot overloads and replace faulty overloads correctly. In this lab you will identify a selection of commonly used overloads in the industry, draw wiring diagrams containing overload circuits, install overloads in circuits, and troubleshoot overloads.

REFERENCES

Sections 9.3 and 11.3

MATERIALS

1. Eight overloads
2. In-line overload
3. Pilot duty overloads
4. Clamp-on ammeter
5. Volt-ohmmeter
6. Miscellaneous wiring supplies

LABORATORY SEQUENCE

A. Overload Identification

1. Obtain eight overloads from your instructor.

2. Record the type of each overload and whether it is in-line or pilot duty.

3. Find and list equipment location of each of the following overloads:
   a. In-line

b. Pilot duty current type

c. Electronic

4. Have your instructor check your work sheet.

B. In-Line Overloads

1. Draw a wiring diagram of an in-line overload protecting a small electric motor.

2. Have instructor check your diagram.

3. Obtain an in-line overload from your instructor and install it on a small electric motor.

4. Operate the motor and record the current draw.

C. Pilot Duty Overloads

1. Complete Lab Figure 14 by connecting the components to form a circuit to protect the motor. Draw a schematic of the circuit.

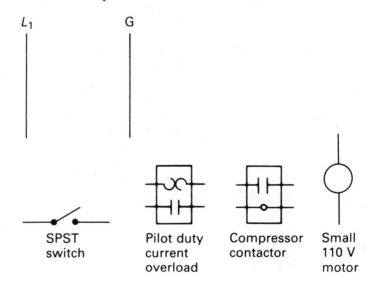

LAB FIGURE 14

2. Have instructor check your diagrams.

3. Install the circuit on an electrical practice board.

4. Have instructor check your wiring.

5. Operate the motor and record the current draw.

6. Remove the components from the practice board and return them to the supply room.

D. Troubleshooting Overloads

Troubleshoot the overloads protecting the compressors of the two air-conditioning units and list the faults.

E. Answer the following questions.

1. Which of the following is not a type of overload used in the industry?
   a. fuse
   b. circuit breaker
   c. transformer
   d. compressor internal thermostat

2. What should you check on a pilot duty overload?
   a. contacts
   b. elements
   c. a and b
   d. none of the above

3. How should you check the element of a pilot duty overload?
   a. Check the resistance of the element.
   b. Check the contacts of the overload.
   c. Check the current draw of the load.
   d. Do all of the above.

4. When troubleshooting an overload, you should note _____.
   a. the current draw of the load
   b. the voltage of the load
   c. the condition of the contacts
   d. all of the above

5. You troubleshoot an overload and see that it opens at a lower amperage than its rated cut-out point. You should _____.
   a. change the overload
   b. change the load
   c. rewire the system
   d. do none of the above

Wiring a Split Air-Conditioning System

PURPOSE

A split air-conditioning system consists of an outside condensing unit and air-moving source plus an evaporator coil located somewhere else. To install this system, you must correctly size the power wiring for the indoor and outdoor sections and make the proper electrical connections. In this lab you will install wiring and check the operation of the equipment.

REFERENCES

Chapters 6 and 12

MATERIALS

1. One air-conditioning system, including condensing unit and evaporator blower. All refrigerant lines should be connected.
2. Volt-ohmmeter
3. Miscellaneous wiring supplies
4. Hand tools
5. Thermostat
6. Disconnect switch

LABORATORY SEQUENCE

A. Wiring a Split Air-Conditioning System

1. Obtain equipment assignment from instructor.

2. Carefully read the installation instructions for the assigned units.

3. Determine and record the wire sizes for the condensing unit and evaporator blower motor.

4. Have instructor check your wire sizes.

5. List the materials required for making all necessary electrical connections. Determine the lengths of the electrical circuits for the power wiring and the control wiring.

6. Have instructor check your list of materials.

7. Install the power circuit from the source to the condensing unit and make the wiring connections. (**Note:** Be careful when making connections in the main electrical panel.)

8. Install the power circuit from the source to the evaporator blower motor and make the wiring connections.

9. Install the thermostat and make all control circuit connections.

10. Have instructor check your installation.

11. Operate the system and record the voltage supplied to each section and the current draw of the compressor, condenser fan motor, and evaporator blower motor.

12. Consult your instructor to determine if the unit is to remain wired or if you should remove the circuits.

B. Answer the following questions.

1. A split air-conditioning system has _____.
   a. a condensing unit
   b. an evaporator
   c. a source for air movement
   d. all of the above

2. When installing an air-conditioning system, you must follow _____.
   a. National Electrical Code
   b. local codes
   c. manufacturer's installation instructions
   d. all of the above

3. The length of an electrical circuit is important because it is possible

to have _____.
   a. voltage drop
   b. lower than normal current draw of the load
   c. high voltage
   d. none of the above

4. The control wiring of a normal split air-conditioning system would be connected to the _____.
   a. thermostat, condensing unit, and power panel
   b. condensing unit, transformer, and indoor fan unit
   c. transformer, thermostat, and indoor fan unit
   d. thermostat, condensing unit, indoor fan unit, and transformer

5. The most important consideration for equipment installation is _____.
   a. conductor insulation
   b. conductor size
   c. disconnect type
   d. none of the above

Wiring a Packaged Air-Conditioning Unit

PURPOSE

A packaged air-conditioning unit is usually made in one piece. It is a popular unit and used throughout the industry. To install this system, you must correctly size the power wiring for the unit and make the proper power and control electrical connections. In this lab you will install the electrical wiring and check the operation of the equipment.

REFERENCES

Chapters 6 and 12

MATERIALS

1. Packaged air-conditioning unit
2. Volt-ohmmeter
3. Miscellaneous wiring supplies
4. Hand tools
5. Thermostat

LABORATORY SEQUENCE

A. Wiring a Packaged Air-Conditioning Unit

    1. Obtain equipment assignment from your instructor.

    2. Carefully read the installation instructions for the assigned unit.

    3. Determine and record the wire size for the packaged unit.

4. Have instructor check your wire size.

5. List the materials required for making all necessary electrical connections. Determine the lengths of the electrical circuits for the power wiring and the control wiring.

6. Have instructor check your list of materials.

7. Obtain electrical supplies from supply room.

8. Install the power circuit from the source to the packaged unit and make the wiring connections.

9. Install the thermostat and make the control circuit connections.

10. Have instructor check your installation.

11. Operate the system and record the voltage supplied to the packaged unit and the current draw of the compressor, condenser fan motor, and evaporator fan motor.

12. Consult your instructor to determine if the unit is to remain wired or if you should remove the circuits.

B. Answer the following questions.

1. A packaged unit is _____.
   a. a self-contained unit
   b. split into four sections
   c. mounted in an attic
   d. none of the above

2. For an installation you must protect _____.
   a. the power supply wire from overload
   b. the control circuit
   c. the indoor fan motor
   d. all of the above

3. The field control wiring on a packaged unit connects _____.
   a. the thermostat and the unit
   b. the thermostat and the transformer
   c. the unit and the transformer
   d. none of the above

4. After installing the unit you should check _____.
   a. the unit voltage
   b. the unit current draw
   c. the compressor current draw
   d. all of the above

5. No installation is complete without _____.
   a. checking the operation of the unit
   b. informing the homeowner of important facts
   c. securely installing unit covers
   d. all of the above

Troubleshooting Air-Conditioning Systems

PURPOSE

Troubleshooting systems can be easy if you use your knowledge of electrical controls and control systems. At least 75% of the problems will be electrical.

REFERENCES

Chapters 11 and 13

MATERIALS

1. Operating air-conditioning systems
2. Volt-ohmmeter
3. Clamp-on ammeter

LABORATORY SEQUENCE

A. Troubleshooting Problems

1. You answer a complaint of "no cooling." The condenser fan motor and evaporator fan are operating. You check the schematic diagram (see Figure 4.38 of the text) and decide the problem is _____.
   a. an open internal overload in the compressor
   b. blown fuses in the disconnect
   c. a bad transformer
   d. a bad thermostat

2. You answer a complaint of "not enough cooling." All unit compo-

nents are operating, but after about 5 minutes, the compressor cuts out. You check the pressures of the refrigeration system and find that the discharge pressure is 160 psig and the suction pressure is 20 psig. The refrigerant in the system is R-22. You check the schematic diagram (see Figure 4.38 of the text) and decide the problem is _____.

a. a bad transformer
b. a bad thermostat
c. an open low-pressure switch
d. an open high-pressure switch

3. You answer a complaint of "no cooling." The compressor and condenser fan motor are operating, but the indoor fan motor is not. (See Figure 4.38.) The probable cause is _____.

a. a bad indoor fan motor
b. a bad transformer
c. a bad indoor fan relay
d. possibly a and c

4. You answer a complaint of "not enough cooling." The compressor, OFM1, and OFM3 are operating. All three OFMs are good. The high-pressure switch opens and stops the compressor. You check the schematic diagram (see Figure 13.08 of the text) and decide the problem is a bad _____.

a. high-pressure switch
b. oil safety switch
c. outdoor thermostat (TS1)
d. fuse

5. You answer a complaint of "no cooling." The only system component operating is the crankcase heater. You turn the fan switch to "on," and the IFR is energized. Checking the schematic diagram (see Figure 13.12), you decide the problem is a bad _____.

a. transformer
b. compressor
c. control relay
d. holding relay

6. You answer a complaint of "no heating." The compressor and outdoor fan motor are not operating but are good. The indoor fan is operating properly. After checking the schematic diagram (see Figure 4.47), you determine that the problem is a bad _____.

    a. transformer
    b. thermostat
    c. MS contactor
    d. supplementary heater

7. You answer a complaint of "not enough heat." The compressor, condenser fan motor, and indoor fan motor are operating, but there is a heavy coating of ice on the outdoor unit coil. The defrost thermostat is closed. (See Figure 4.47 for the schematic diagram.) The probable cause is _____.
    a. a bad defrost timer
    b. a bad transformer
    c. a bad defrost relay
    d. possibly a and c

8. You answer a complaint of "no cooling." The compressor is not operating, but the condenser fan motor and indoor fan motor are. The compressor body is hot. (See Figure 13.3 for the schematic diagram.) The probable cause is a _____.
    a. bad contactor coil
    b. bad running capacitor
    c. blown fuse
    d. bad transformer

B. Troubleshooting Heating, Air-Conditioning, or Refrigeration Systems

1. Your instructor will assign you three systems to troubleshoot.

2. Troubleshoot system 1.

3. Repair system 1.

4. Have instructor check unit 1.

5. Troubleshoot system 2.

6. Repair system 2.

7. Have instructor check unit 2.

8. Troubleshoot system 3.

9. Repair system 3.

10. Have instructor check unit 3.

C. Answer the following questions.

1. When troubleshooting a unit, you should first _____.
   a. make a visual inspection of the unit
   b. check the transformer
   c. check the compressor
   d. check the charge

2. When troubleshooting a unit that is not operating, the correct sequence of checks is _____.
   a. thermostat setting, power supply to unit, transformer
   b. transformer, power supply to unit, thermostat setting
   c. thermostat setting, charge of unit, power supply to unit
   d. compressor, power supply, thermostat setting

3. Which of the following items are essential for troubleshooting a unit?
   a. good electrical meters
   b. understanding schematic wiring diagrams
   c. knowing how the unit works
   d. all of the above

4. When troubleshooting a compressor motor, you must check the _____.
   a. transformer
   b. run capacitor
   c. system discharge pressure
   d. system suction pressure

5. You must have the right attitude toward _____.
   a. the homeowner
   b. your employer
   c. fellow employees
   d. all of the above

Glossary

air-cooled packaged unit: A unit that is made in one complete unit with an air-cooled condenser mounted with the unit or remotely.

alternating current: Electron flow that flows in one direction and then reverses at regular intervals; produced by cutting a magnetic field with a conductor. The most common type of power supply used in the heating, cooling, and refrigeration industry.

ambient air temperature: The temperature of the air surrounding any device.

ammeter: An electric meter used to measure the amperes that are present in a circuit. Ammeters are made as an in-line ammeter or clamp-on meter.

ampere: The amount of current required to flow through a resistance of one ohm with a pressure of one volt, measured with an ammeter.

anticipator: A component of a thermostat that anticipates the temperature of an area and will stop or start the cooling or heating equipment to prevent the thermostat from overshooting the desired temperature.

anti-short-cycling device: A device that prevents a load from stopping and starting in rapid succession. This device is used mainly to prevent compressors from stopping and starting rapidly.

apparent wattage: The power that is calculated by multiplying the voltage times the amperage of a circuit.

armature: The portion of a contactor that moves; connected to a set of contacts that causes a completed circuit when the armature is pulled into the magnetic field produced by the coil.

atom: The smallest particle of an element that can exist and maintain any identification; can combine with other atoms to form new substances.

back electromotive force: The amount of voltage produced across the starting and common terminals or connections of a single-phase motor.

ball bearing: An antifriction device that is used to allow free turning and support of the rotating member of the device. It consists of an outer ring and inner ring with races and has steel balls sandwiched between the rings.

bearing: The part of a rotating machine or motor that allows free turning of its rotating parts with little friction.

bimetal overload: A simple thermal overload that breaks the power supplying a small motor directly. When it reaches a high enough temperature, the bimetal opens by warping; when the overload cools, the bimetal will warp back and close the circuit.

bimetal relay: An overload device that opens a set of contacts by a temperature that corresponds to current draw of a load.

breaker: A device usually used in a breaker panel that is capable of being used as a disconnect switch and an overload for the circuit it is supplying power to.

breaker panel: An electric panel that houses breakers used to distribute power to circuits in the structure.

callback: A return trip to a piece of equipment that was not properly serviced the first trip.

capacitive reactance: A resistance caused by using capacitors with motors; when a capacitor is put in a circuit, it resists the voltage change, causing the amperage to lead the voltage.

capacitor: A device that consists of two aluminum plates with an insulator between the plates; used to boost starting torque of single-phase motors. The two types of capacitors are electrolytic or starting and running or oil-filled.

capacitor-start-capacitor-run motor: A high starting torque and good running efficiency motor; it uses a starting capacitor to increase starting torque and a running capacitor to increase running efficiency.

capacitor-start motor: A motor that uses a starting capacitor to increase the starting torque.

charge: A condition in which an imbalance of protons and electrons exists in an atom.

closed circuit: A complete path for electrons to follow.

compressor: A device that is used in a mechanical refrigeration system to compress the refrigerant. Most compressors are rotated by an electric motor. The motor may be external or be an integral part of the compressor.

condensing unit: A portion of a split air-conditioning or refrigeration system that is mounted outside and contains the compressor, condenser, condenser fan motor, and controls for these components; most used today are air-cooled. It takes the cool suction gas from the evaporator, compresses it, condenses it to a liquid, and forces it back to the evaporator.

conductor: A wire that is used for the path of electric flow. Most electric conductors are copper or aluminum.

contactor: A device that opens and closes a set of electric contacts by the action of a solenoid coil; composed of a solenoid and the contacts.

continuity: A complete path for electrons to follow in a circuit or component.

control circuit: A circuit that controls some load in the entire control system, whether it be a relay or contactor coil or a major load.

current: Electrons flowing in an electric circuit, measured in amperes.

current electricity: Electricity that results from the electron being displaced and moving back to the atom.

current overload: An overload that opens a set of contacts on high current draw and allows them to close when the current draw has decreased. It usually is a pilot duty device.

current relay: A relay that is opened or closed by the starting current of an electric motor. The relay allows a starting capacitor and starting winding to drop out or drop in the starting circuit.

cut-in pressure: The pressure at which a pressure control will close, starting the device it is controlling.

cut-out pressure: The pressure at which a pressure control will open, stopping the device it is controlling.

cycle: One complete cycle of alternating current is the production of a positive and negative peak.

deenergize: To stop the electron flow to an electric device.

defrost control: A control that is used to initiate the defrost cycle of a heat pump or initiate the defrost in a commercial refrigeration freezer.

delta transformer: A three-phase transformer that has the ends of each of its windings connected together to form a triangle. The delta transformer produces a high leg on one of its power legs.

delta winding: A winding layout of some three-phase motors, where the beginning of the windings is connected to the ending of the windings.

dielectric: The substance that is between the plates and fills the case of a capacitor. The substance is a nonconductor of electricity.

differential: The difference between the cut-in and cut-out point of a control. This can be applied to thermostats, pressure switches, and most controls.

diode: A semiconductor that allows voltage to pass in only one direction.

direct current: An electron flow in only one direction; used in the industry only for special applications such as solid-state modules, electronic air filters, and special applications.

direct drive: A method of transferring rotating motion of a motor to a device that must be turned. This type of hookup connects the motor directly to the device that must be turned and rotates the same revolutions per minute as the motor.

disconnect switch: A switch that is used to disconnect the power supply to a piece of equipment; it is sometimes referred to as a safety switch.

distribution center: A electric panel used to distribute electric supply to several places in a large structure; can be of fusible or circuit breaker design.

effective voltage: Alternating current, with its many reversals and peaks, never peaks at a constant value. The effective voltage is the working voltage of alternating current. The effective voltage is 0.707 times the peak voltage.

electric circuit: A path for electrons to follow; the circuit may be open or closed depending on the position of its switches.

electric energy: Energy that is produced by a movement of electrons. The energy can be produced by chemical, light, thermal, or mechanical means.

electricity: Energy that is capable of producing an electron flow. An unbalanced condition that results when an electron can be easily displaced from an atom.

electric meter: A device used to measure some electrical characteristics of a circuit such as the voltage, amperage, resistance, or wattage.

electric power: The rate at which electricity is being used, measured in watts.

electric pressure: Another term used to refer to electromotive force, potential difference, and voltage.

electric switch: A device that opens or closes to control some load in an electric circuit. It can be opened or closed by temperature, pressure, humidity, flow, and manual means.

electrodes: The dissimilar metal conductors in a battery that produce a small difference of potential.

electrolyte: The chemical paste between the electrodes of a battery and some capacitors.

electromagnet: A magnet produced by coiling wire around a metal core.

electromotive force (emf): The difference of potential that forces electrons through a resistance.

electron: Particles that orbit around the nucleus of an atom and have a negative charge.

element: A substance that has weight, takes up space, and cannot be broken down by chemical means.

energize: To apply voltage to an electric device.

energy efficiency rate (EER): The means of measuring an air conditioner for its efficiency by stating how many Btu's of cooling are available from one watt.

factory-installed wiring: The wiring installed in a piece of equipment at the factory; usually the connections between the components in the control panel and the system components in the unit itself.

factual diagram: A wiring diagram that is a combination of the pictorial and schematic diagrams.

field of force: The area around a magnet that is affected by the strength of the magnet.

field wiring: The wiring that must be installed in the field by the installation mechanic.

flow diagram: A block diagram that outlines the operation of a heating and air-conditioning unit.

flux: The magnetic lines of force of a magnet that connect the north and south poles of the magnet.

free electron: Electrons that are easily removed from the outer orbits of atoms.

frequency: The number of complete cycles per second of alternating current.

fuse: A device that breaks a circuit when its ampere rating is exceeded; constructed of two ends or conductors with a piece of wire that will melt and break the circuit on an overload.

fusible load center: An electric panel that supplies circuits with power and protects them with fuses.

gas pack: A unit that heats in the winter by using gas as its fuel and cools in the summer by using electric power.

head pressure: The discharge pressure of a refrigeration system, sometimes called high-side pressure.

heater: An electric load that converts electric energy to heat.

heat pump: A refrigeration system that reverses the flow of refrigerant in the normal refrigeration cycle, which allows the unit to cool in the summer and heat in the winter.

hermetic compressor motor: A motor that is designed for single- and three-phase operation and is totally enclosed in a shell with refrigerant and oil.

hertz: The number of complete cycles per second of alternating current; more widely accepted than the term "frequency."

hopscotching: A troubleshooting procedure for electric circuits that is accomplished by jumping from one component to another.

hot-wire relay: A relay that is opened or closed by a thermal element that senses the starting current of the motor. The relay allows a starting capacitor or starting winding to drop out or drop in the starting circuit. This type of relay also has a built-in means of overload protection.

humidistat: A device that is used to control humidity; it uses a moisture-sensitive element to control a mechanical linkage that opens and closes an electric switch.

impedance: The sum of the resistance and reactance in an alternating current circuit.

indoor fan relay: An electric relay that starts and stops an indoor fan on cooling, electric-heating, and heat pump systems.

induced magnetism: The magnetism induced around a current-carrying conductor.

inductance: A property of an alternating current circuit by which an electromotive force is produced in it by a variance in current.

inductive load: A load that starts with a larger ampere draw and reduces it as the load starts normal operation. The increase in the ampere draw initially is due to inductance.

inductive reactance: The opposition to the change in alternating current flow that produces an out-of-phase condition between voltage and amperage.

installation diagram: A diagram that shows little internal wiring but gives specific information as to terminals, wire sizes, color coding, and breaker or fuse sizes.

insulator: A material that retards the flow of electrons or electricity.

internal compressor overload: An overload that is embedded in the windings of a motor. Some internal overloads break the power to the motor directly while others merely open a set of contacts that is wired into an electric control circuit.

kilowatthour: The rate at which electric energy is being used at a specific time. Most electric utilities bill their customers in this method.

law of electric charges: Like charges repel and unlike charges attract.

line break overload: An overload that breaks the power going to the motor and is most commonly used on small motors.

line voltage: The voltage being supplied to the equipment at the power supply.

load: Electric devices that consume electricity to do useful work, such as motors, solenoids, heaters, and lights.

load analyzer: A device that analyzes the load of a structure and determines the correct action of a heating and cooling unit.

locked rotor amperes: The current a motor uses the instant it starts while the rotor is in a stationary position.

magnetic field: The area around a magnet in which the effect of magnet can be felt.

magnetic overload: An overload device that senses the current draw of a load by the magnetic field produced, which is proportional to the current draw. The device will open a set of contacts on high current draws and allow them to close when the ampere draw returns to normal.

magnetic starter: A device that opens and closes its contacts when a solenoid is energized. A means of overload protection is provided. It is the same as a contactor except for the overload protection.

magnetism: The ability of two pieces of iron to be attracted to each other by physical means or electrical means.

matter: The substance of which all physical objects consist.

measurable resistance: The actual resistance of a circuit or component measured with an ohmmeter.

microfarad: The unit of measurement used to measure the strength of a capacitor.

module: An electrical device that is used to control one function or more of a control system.

molecule: The smallest particle into which an object can be divided and maintain the properties of that object.

molten-alloy relay: An overload device that opens a set of contacts by thermal energy. This type of device allows the temperature produced by the starting current of a load to be transferred to a molten-alloy device. When it reaches a certain temperature, it will melt the solder around the device, causing it to slip and open the contacts; when it cools it will harden again and the relay must be manually reset.

motor: A device used to create a rotating motion and drive components that require rotating motion. Electric energy is changed to mechanical energy by magnetism, which causes the motor to turn.

multizone: A heating and cooling unit that is equipped to condition more than one zone in a structure.

negative charge: The result of electrons joining atoms.

neutron: The neutral particle in the nucleus of an atom.

noninductive load: A load that has only resistive qualities with no inductive qualities. An electric heater and incandescent lighting are two common types of noninductive loads.

normally closed: The position of a set of contacts or other electric devices that are closed when the device is deenergized.

normally open: The position of a set of contacts or other electric devices that are open when the device is deenergized.

nucleus: The central part of an atom composed of protons and neutrons.

ohm: The amount of resistance that will allow one ampere to flow with pressure of one volt.

ohmmeter: An electrical meter used to measure the resistance of a circuit or electric component.

Ohm's law: The relationship between current, electromotive force, and resistance in an electric circuit.

open circuit: A circuit without a complete path for electrons to follow.

out of phase: The voltage and current are not working together.

overload: A device that is used to detect a high ampere draw of some electric load and

break the controlling circuit, stopping the load.

overshoot: The additional heating or cooling that has been delivered to the conditioned space after the thermostat contacts have opened.

packaged air-conditioning unit: A system built with all components in one unit except for the field wiring.

parallel circuit: An electric circuit that has more than one path for current flow.

peak voltage: When the voltage reaches its peak in an alternating current circuit.

permanent magnet: A piece of material that is magnetic by physical means. Iron, nickel, cobalt, and chromium are materials that can easily be magnetized and will maintain their magnetism for a period of time.

permanent split-capacitor motor: An electric motor, widely used in the industry, that has a moderate starting torque and good running efficiency.

phase: The number of currents alternating at different times in an alternating current circuit.

pictorial diagram: A wiring diagram that shows the actual internal wiring of a unit, much like a picture taken of a control panel. It is also called a line or label diagram.

pilot duty: A term used to refer to an electric device that indirectly controls a major load because of its large ampere draw but controls it directly through a device that is capable of carrying the load.

pole: One set of electric contacts either in an automatic device or a manual switch. Electric devices such as relays, contactors, switches, and breakers can be purchased with one or many poles.

positive charge: The result of electrons leaving an atom.

potential coil: A coil energized by a voltage being applied to it. It can be designed to operate on 24, 110, 208/230, or 460 volts. These coils are used on relays, contactors, and magnetic starters.

potential difference: Two points that have a difference in electric charge; the electric difference between two points in an electric circuit.

potential relay: A relay that uses the back electromotive force of a motor to drop out the starting apparatus when the motor reaches 75% of full speed.

power factor: The ratio of true power to apparent power, usually expressed as a percentage.

pressure switch: A device that opens or closes a set of contacts when a certain pressure is applied to the diaphragm of the switch.

proton: A positively charged particle in the nucleus of an atom.

pump-down control system: A control system that closes a solenoid to allow the compressor to pump all the refrigerant from the low side of the system into the high side. This system is used on large air-conditioning systems and some commercial refrigeration systems.

push-button switch: A switch that can be opened or closed by pressing buttons on the switch. Push-button switches come with a wide variety of purposes and labeling.

range: The operating ranges or limits of a control.

reactance: The resistance that alternating current encounters when it changes flow.

rectifier: A device that will allow electrical current to pass in one direction but stops electrical current from flowing in the opposite direction. This device is commonly used to rectify AC voltage to DC voltage.

relay: A device that opens and closes a set of contacts when its coil is energized. The relay is much like the contactor except for its smaller size.

relief valve: A device that will open on a rise in pressure and release pressure to return a closed system to a safe operating condition and close when the pressure has decreased.

reset point: The point at which an electrical control will close its contacts after an unsafe condition has corrected itself.

resistance: The opposition to the flow of electrons.

rooftop unit: A heating and cooling unit that conditions a structure; it is mounted on the roof after adequate reinforcement has been built into the roof.

rotor: The rotating part of an electric motor.

running (oil-filled) capacitor: An electric device that is used to momentarily store electrons and create a second phase in the starting winding circuits of single-phase motors. This type of capacitor is designed to stay in the circuit whenever the motor is running as a means of heat dissipation.

safety device: Any device that is in a control system for the purpose of safer operation of a major load.

schematic diagram: A diagram that lays out the control system circuit by circuit and is composed of symbols representing components and lines representing their interconnecting wiring.

semiconductor: A conductor whose electrical conductivity is between an insulator and a metal.

sensor: A device that produces a signal that changes with a temperature change.

series circuit: An electric circuit that has only one path for electron flow.

series-parallel circuit: A combination of series and parallel circuits.

shaded-pole motor: An induction type of motor that does not incorporate an ordinary type of starting winding. It uses a band on one side to obtain a short-circuit effect that produces a rotating magnetic field. This motor has a low starting torque.

short circuit: An electric circuit that has no resistance.

short cycling: A term used to refer to a condition that occurs when a load is stopping and starting too frequently.

signal light: A light that is used to show when some electric component or circuit is energized by illuminating the light.

sine wave: A graphical representation of alternating current; a graph showing the sine function of all angles from 0 to 360 degrees.

sleeve bearing: An antifriction device that allows free turning and support of the rotating member of a device. It consists of a solid piece of bronze or babbit that is round and drilled to the diameter of the shaft. The bearing is sometimes called a plain bearing or bushing.

sliding armature: An armature that mounts between two slots in a contactor frame and moves up and down the slots when the contactor is energized.

snap action of a thermostat: The closing of a set of contacts of a thermostat with a snapping motion rather than with a light contact.

solenoid: A device that, when energized, will create a magnetic field and cause some action to an electric component. It opens and closes to control some element of a heating, cooling, and refrigeration control system.

solenoid valve: A valve that opens or closes by a solenoid coil being energized to pull a steel core into the magnetic field of the solenoid.

split-phase motor: An electric motor that has a running and starting winding. This is an induction type of motor.

squirrel cage rotor: The rotating part of an electric motor; its name is derived from its appearance of a squirrel cage. This type of rotor is used in split-phase, capacitor-start, shaded-pole, and three-phase motors.

staged system: A system that has more than one mode of heating or cooling operation.

staging thermostat: A thermostat that is designed to open and close more than one set of contacts to control several modes of heating or cooling operation.

starting capacitor: An electric device that is used to momentarily store electrons, creating a second phase in starting windings of single-phase motors. This type of capacitor is designed to stay in the circuit only a short period of time.

starting relay: A relay that is used to energize or deenergize the starting components of a single-phase motor.

star transformer: A three-phase transformer that has the ends of each winding connected to a common point. The star transformer produces a balance of all hot legs to ground.

star winding: A winding layout of some three-phase motors in which the end of the windings are connected together.

static electricity: Electricity that results from the electron being displaced and not returning to the original atom; usually results from friction.

stator: The stationary part of an electric motor.

swinging armature: An armature used in a contactor that is mounted on a line and moves up and down in a swinging motion.

switch: A device for making, breaking, or changing the connection in an electric circuit.

system lag: The difference in temperature between the point at which the thermostat closes and the point at which the thermostat starts to rise or fall.

thermal overload: An overload device that senses the current draw of a load by the heat produced, which is proportional to the current draw.

thermal relay: An overload device that determines current flow and opens a set of pilot duty contacts when an overload is indicated.

thermistor: A semiconductor that has a temperature coefficient of resistance that corresponds with a designated temperature. This device is widely used as a signal device in a control circuit.

thermostat: A device that responds to a temperature change by opening or closing a set of electric contacts.

thermostat controlling element: The portion of a thermostat that reacts to temperature change by opening or closing the contacts through mechanical linkage. The two types of elements are bimetal and bulb.

three-phase motor: An induction type of motor that has a very high starting torque and requires no special starting apparatus. The motor must be operated on three-phase current.

throw: This refers to the number of positions of the movable contacts that will complete a circuit.

torque: The starting power of an electric motor.

transformer: A device that decreases or increases the incoming voltage to the desired voltage.

transistor: A semiconductor device used for the control and amplification of a signal from one circuit to another circuit.

trip-out point: The point at which an electrical control will open its contacts in the event of an unsafe condition.

varistor: A semiconductor with a voltage-sensitive resistance.

V-belt: The belt that connects the pulleys of a motor and the device that must be rotated and transfers the rotating motion from the motor to the device. V-belts can be purchased in several widths and almost any length.

volt: The amount of electric pressure required to force one ampere through a resistance of one ohm.

voltage: The difference in electric potential between two points.

voltage drop: The amount of voltage lost through any type of switching device or conductor.

voltmeter: An electric meter used to measure voltage.

water chillers: A refrigeration system that cools water that is pumped into other parts of the system to maintain the desired condition in a specific area.

water-cooled packaged unit: A unit that is made in one complete unit with a water-cooled condenser as an integral part.

watt: One ampere flowing with a pressure of one volt. The unit measurement of power.

wiring diagram: A systematic method of laying out the wiring that is interconnecting the control components within the control system; three types are schematic, pictorial or line, and installation.

Y transformer: A three-phase transformer that has a common junction point and forms a Y. This transformer hookup allows for a completely balanced load when using all hot legs and ground.

zone: A section of a structure that has a different heating and cooling load.

Index